Economy-Wide Policy Modeling of the Food-Energy-Water Nexus

DEVELOPMENT ECONOMICS AND POLICY

Series edited by Franz Heidhues †, Joachim von Braun, Ulrike Grote and Manfred Zeller

Vol. 79

PETER LANG

Economy-Wide Policy Modeling of the Food-Energy-Water Nexus

Identifying Synergies and Tradeoffs on Food, Energy, and Water Security in Malawi

Franziska Schünemann

PETER LANG

Bibliographic Information published by the Deutsche Nationalbibliothek
The Deutsche Nationalbibliothek lists this publication in the Deutsche Nationalbibliografie; detailed bibliographic data is available in the internet at http://dnb.d-nb.de.

Library of Congress Cataloging-in-Publication Data
A CIP catalog record for this book has been applied for at the Library of Congress.

Zugl.: Hohenheim, Univ., Diss., 2017

D 100
ISSN 0948-1338
ISBN 978-3-631-75618-8 (Print)
E-ISBN 978-3-631-76110-6 (E-Book)
E-ISBN 978-3-631-76111-3 (EPUB)
E-ISBN 978-3-631-76112-0 (MOBI)
DOI 10.3726/b14357

Peter Lang – Berlin · Bern · Bruxelles · New York · Oxford · Warszawa · Wien

This publication has been peer reviewed.

www.peterlang.com

Table of Contents

Acknowledgments 11

Abstract 13

List of abbreviations 17

List of tables 19

List of figures 21

1 Introduction 23

1.1 Background: global challenges and the food-energy-water nexus 23

1.2 The FEW nexus in the broader debate on sustainable development 28

1.3 The role of linkages in the FEW nexus in detail: economic and ecological linkages 30

1.4 Objectives and research questions 32

1.5 Policy measures analyzed and specific research questions 33

1.6 Structure of the study 36

References 37

2 Developing the appropriate modeling framework for quantitative policy analysis of the food-energy-water nexus 41

2.1 The importance of economic linkages 42

2.2 Economic policy simulation models and their applicability to analyze the nexus 45

2.2.1 Farm household models 46

2.2.2 Partial equilibrium models 48

2.2.3 Computable general equilibrium models 51

2.2.4 CGE models' suitability for analyzing the FEW nexus in Malawi and the specific challenges ... 57
2.2.4.1 Food ... 58
2.2.4.2 Energy ... 59
2.2.4.3 Water ... 61
2.2.4.4 Conclusion ... 62
2.3 Modeling the nexus: integrating the environment and non-market interactions ... 63
2.3.1 Integrated global models ... 64
2.3.2 An integrated modeling framework for nexus analysis in Malawi ... 65
2.3.2.1 Malawi CGE model ... 65
2.3.2.2 Modeling framework for analyzing biofuel production ... 70
2.3.2.3 Modeling framework for analyzing irrigation expansion ... 72
2.3.2.4 Modeling framework for analyzing the biomass energy sector ... 74
2.4 Conclusion ... 76
References ... 77
3 Leveling the field for biofuels: comparing the economic and environmental impacts of biofuel and other export crops in Malawi ... 83
3.1 Introduction ... 83
3.2 Biofuels in Malawi ... 85
3.2.1 Food, energy, and water systems ... 85
3.2.2 Sugarcane-ethanol ... 86
3.2.3 Ethanol production technologies ... 87
3.3 Measuring economic and environmental impacts ... 89
3.3.1 Measuring economy-wide impacts ... 89
3.3.2 Simulating biofuels production ... 91

3.3.3 Estimating crop water use 93
3.3.4 Measuring greenhouse gas emissions 94
3.4 Results 95
3.4.1 Baseline scenario 95
3.4.2 Producing biofuels on estate farms 97
3.4.3 Using outgrower schemes 99
3.4.4 Relying on rain-fed production 100
3.4.5 Environmental impacts and trade-offs 101
3.4.6 Imposing stricter land constraints 104
3.4.7 Biofuels versus other export crops 105
3.5 Conclusion 108
References 109
Appendix 112
4 Evaluating irrigation investments in Malawi: economy-wide impacts under uncertainty and labor constraints 117
4.1 Introduction 117
4.2 Agriculture and irrigation in Malawi 119
4.2.1 Irrigation master plan 121
4.3 Modeling approach 122
4.3.1 Crop models 122
4.3.2 Economy-wide model 124
4.3.3 Simulating irrigation expansion 126
4.3.4 Simulations 127
4.4 Simulation results 130
4.4.1 Irrigating summer crops 132
4.4.2 Introducing a winter season 137
4.4.3 Increasing irrigation with unemployment 138
4.4.4 Accounting for climate uncertainty 139
4.4.5 Irrigation fertilizer interaction 141

4.5 Conclusion 143
References 144
Appendix 147

5 Policies for a sustainable biomass energy sector in Malawi: enhancing energy and food security simultaneously 149

5.1 Introduction 149
5.2 Background: energy sector development in Malawi 151
5.2.1 Population growth, income growth and urbanization 152
5.3 Estimating biomass energy demand 154
5.3.1 Estimating household demand 154
5.3.2 Household demand projections 156
5.3.3 Non household demand 158
5.3.4 Results and discussion 159
5.4 Estimating biomass supply 162
5.4.1 Results and discussion 163
5.5 Demand side policy: improved cookstoves 165
5.5.1 Results and discussion 166
5.6 Supply side policy: agroforestry 168
5.6.1 Methodology 168
5.6.2 Results and discussion 172
5.7 Conclusion and policy implications 176
References 178
Annex 183

6 Discussion and conclusions 187

6.1 Major results and achievements 187
6.1.1 Methodological contributions 187
6.1.2 Empirical findings 190
6.1.2.1 Simultaneous impacts on food, energy, and water security 190

6.1.2.2 Role of drivers 192
6.1.2.3 Impact on the livelihoods of the poorest 193

6.2 Policy implications 195
6.2.1 Smallholder integration 195
6.2.2 Funding of policies 196
6.2.3 The food-energy-water nexus as a paradigm for the Sustainable Development Goals 196
6.2.4 Traditional fuels for sustainable development 197

6.3 Directions for further research 198

References 200

General bibliography 203

Acknowledgments

This study emerged during my work in the project "Policies and Institutions for Achieving the Virtuous Food-Energy-Water Nexus in Sub-Saharan Africa" of the International Food Policy Research Institute (IFPRI) and the University of Hohenheim. First of all, I would to like thank Professor Manfred Zeller for giving me the chance to become a part of such an exciting project and to advance in my research skills. From the beginning, he was confident that I would be able to work into a completely new area of research, and guided and supported my research ideas from Germany and Uganda. I am deeply indebted to James Thurlow and Siwa Msangi of IFPRI who have been my mentors on economy-wide and natural resource modeling. They have been supporting me throughout my PhD to be innovative and put my ideas into mathematical code. I spent a considerable amount of time doing research at IFPRI in Washington DC where I became part of the Environment and Production Technology Division. I would like to express my gratitude to Ephraim Nkonya and his wife Leticia for their warm welcome into their family. I would like to thank Karl Pauw and Tingju Zhu for sharing with me their knowledge in social accounting matrices and water modeling. I would like to thank my coauthors Joao Rodrigues and Stefan Meyer for providing me with the latest information from Malawi. I am extremely grateful to all Malawian stakeholders that took part in several project workshops in Lilongwe and provided my with invaluable insights on food, energy, and water security in Malawi.

My time in Hohenheim was made very special and memorable by a group of great colleagues, who have become real friends. I will miss you all dearly and am very thankful for the research discussions and leisure you shared with me.

This work is dedicated to my beloved parents. Their continuous support and love have given me the strength to aspire to my dreams and to follow in their footsteps. I would also like to thank my siblings and their families as well as my cousin for being always there for me. I would like to thank my American parents for giving me a real home in the United States. To my fellow sisters, thanks for always helping with words and deeds. Last but not least, I would like to thank my partner for his unconditional love and belief in me.

Finally, I would like to gratefully acknowledge funding by the German Federal Ministry for Economic Cooperation and Development (BMZ) for the CGIAR research project entitled "Policies and Institutions for Achieving the Virtuous Food-Energy-Water Nexus in Sub-Saharan Africa".

Abstract

Food, energy, and water are essential goods for human survival and are intrinsically connected through economic linkages and ecological processes. All three are dependent on limited resources which are threatened by global drivers in the form of economic growth, population growth, and climate change that are particularly affecting developing countries. In the light of these challenges, researchers and policy makers gathered in Bonn, Germany, in 2011 and agreed that development policy cannot continue on its current "silo" path, but must undergo a transformation towards a nexus perspective of integrated food, energy, and water security policies. This study contributes twofold to the research on the food-energy-water (FEW) nexus: methodologically, through developing integrated modeling frameworks for ex-ante policy assessments that capture the linkages between food, energy, and water, and empirically through identifying those policy measures that maximize the synergies for food, energy, and water (FEW) security and minimize the tradeoffs. To this end, three studies analyze four policies in Malawi – biofuels production, irrigation expansion, improved cookstoves and agroforestry – that directly affect FEW security and provide a large scope for realizing synergies.

Previous studies on FEW security are sector-specific and are unable to simultaneously measure policy effects on all three sectors, which is pivotal to find the most beneficial interventions for the whole FEW nexus. Due to the complexity of the nexus, there is to date no existing economy-wide modeling framework that adequately captures all various interconnections. To close these research gaps three innovative modeling approaches for ex-ante policy simulation are developed that encompass all relevant linkages of the FEW nexus at country level. A computable general equilibrium (CGE) model of Malawi is at the heart of each modeling framework. The economy-wide model includes all economic linkages and the social dimension of the FEW nexus. The CGE model encompasses both the availability dimension of FEW security by measuring output changes and the access dimension through income and prices. Population and economic growth are incorporated within the CGE model to determine the role of these drivers. As the CGE model includes neither nexus linkages outside of market interactions nor ecological processes, the model is linked with biophysical and tailor-made farm household models as suitable for each policy analyzed. Each of these modeling frameworks does not aim to cover all linkages between food, energy, and water, rather the frameworks concentrate on the nexus linkages affected by the policy measures analyzed to better understand the effectiveness of those policies in evoking synergies and tradeoffs. The modeling frameworks therefore close an important methodological gap in the literature by allowing detailed ex-ante policy analysis of the FEW nexus at country level, especially since all developed models can easily be transferred to other developing countries. By simulating the abovementioned four policy measures within these modeling frameworks, their empirical impacts on the FEW nexus in Malawi are determined.

Biofuels are promoted to increase energy security, but often raise the specter of food insecurity, water resource depletion, and greenhouse gas (GHG) emissions from land clearing. These concerns underpin the "sustainability criteria" governing access to European biofuel markets, although it is unclear if producing biofuels in low-income countries like Malawi exacerbates food insecurity or conversely increases income of smallholders. The results of the policy simulations demonstrate that biofuel production reduces poverty and food insecurity in Malawi by raising household incomes. An expansion of biofuel production in Malawi leads to economic growth and higher energy security by increasing the availability of fuel. This synergy between food and energy security is dependent on irrigation of feedstock, which entails tradeoffs for water security and climate change. The high GHG emissions of irrigated feedstock production and biofuel processing do not meet the EU sustainability criteria. Biofuels therefore involve tradeoffs for the environment that need to be valued carefully. In contrast, the findings also reveal that biofuel crops are not worse but even better for the environment and food security than other export crops, making the sustainability criteria overly biased against biofuels.

As irrigation determines the effectiveness of biofuel production, the second study examines impacts of irrigation expansion on the FEW nexus in Malawi. Irrigation is crucial to increase food security and mitigate effects from climate change, but the low profitability has led to little irrigation investments in Sub-Saharan Africa so far. Since benefits from irrigation arise not only from yield increases, but also from multiplier effects and lower climate risks, the developed modeling framework simultaneously assesses the various returns. Water availability is explicitly included in the simulations to minimize tradeoffs for water security. As irrigation is conducted with smallholder technologies such as treadle pumps and watering cans, energy needs are minimal. The policy simulations confirm the low profitability of irrigation in Malawi due to relatively low yields and high labor requirements. Despite that, the study reveals large benefits of irrigation in Malawi through higher food security, lower poverty, and reduced vulnerability to weather variability, making investments in irrigation worthwhile to improve the livelihoods of the poorest. If implemented correctly, irrigation expansion in Malawi is a policy that exhibits only synergies for FEW security.

Biomass energy in the form of firewood and charcoal dominates the energy sector in Sub-Saharan Africa, in particular as the main source of cooking fuel. The strong linkages to food security and the environment make biomass energy crucial for sustainable development. As population and economic growth are exacerbating already existing supply-demand imbalances, the third study analyzes the impact of improved cookstoves and agroforestry on the FEW nexus with a focus on establishing a sustainable biomass energy sector in Malawi. Water security is only indirectly, but positively affected by these policies as lower demand for, and higher supply of trees increases the ability of the soil to hold water. The model results suggest that improved cookstoves and agroforestry have the potential to increase energy security in Malawi in a way that is sustainable for the biomass energy sector. Especially agroforestry increases both availability of and access to energy. As food security is dependent on the secure provision of cooking energy, an increase in energy security simultaneously increases the access to and the utilization of food. The strong link between energy and food security is therefore strengthened by both interventions, leading to a win-win situation for all nexus sectors. Biomass energy can be inherently sustainable and should be an integral part of energy sector strategies in developing countries and the Sustainable Development Goals.

The empirical findings of this study show that policy measures indeed produce some tradeoffs between FEW security, but that – if policies are designed correctly – the tradeoffs can be minimized while simultaneously maximizing the synergies. These findings are an essential contribution to the literature through demonstrating that even in a world with enormous pressures on limited resources, prudent policy making can ensure FEW security for all people. The design of policy measures plays a crucial role for their success and the implementation of these policies will require substantial efforts in terms of funding and integrating smallholders. The policy analyses also highlight that without considering the impact of drivers such as population growth and climate change, policy measures might not be successful and entail undesirable tradeoffs. Finally, the results demonstrate that the development of integrated modeling frameworks is vital for quantitative analyses of policies that simultaneously affect the economic, social, and environmental spheres to identify the synergies and tradeoffs. This study makes an important methodological contribution to integrated environmental-economic modeling of developing countries and may serve as a starting point for future research on linking the economy and the environment in models.

List of abbreviations

AEZ	Agro-Ecological Zoning
CAGR	Compound annual growth rate GOM
CAPRI	Common Agricultural Policy Regionalised Impact
CES	Constant elasticity of substitution
CGE	Computable general equilibrium
CLEW	Climate, Land, Energy and Water
CO_2eq	Carbon dioxide equivalent
DSGE	Dynamic stochastic general equilibrium
DSSAT	Decision Support Software for Agrotechnology Transfer
EBCR	Economy-wide benefit-cost-ratio
EC	European Commission
ESIM	European Simulation Model
ESM	Earth System Models
EU	European Union
EX-ACT	Ex-ante carbon balance tool
FAO	Food and Agriculture Organization of the United Nations
FEW	Food-energy-water/Food, energy and water
FHM	Farm household model
FISP	Fertilizer input subsidy program
GDP	Gross Domestic Product
GHG	Greenhouse Gas
GoM	Government of Malawi
GTAP-W	Global Trade Analysis Project – Water
IAM	Integrated Assessment Models
ICS	Improved cookstoves
ICSU	International Council for Science
IEA	International Energy Agency
IFPRI	International Food Policy Research Institute
IHS	Integrated Household Survey
IIASA	International Institute for Applied Systems Analysis
ILUC	Indirect land use change
IMPACT	International Model for Policy Analysis of Agricultural Commodities and Trade
ISSC	International Social Science Council
LEAP	Long-range Energy Alternatives Planning

LES	Linear expenditure system
LPG	Liquid petroleum gas
LUC	Land use change
MDGs	Millennium Development Goals
MESSAGE	Model for Energy Supply Strategy Alternatives and their General Environmental Impact
MJ	Mega joule
MKW	Malawian Kwacha
NPR	Net present value
NSO	National Statistical Office
PDF	Probability density functions
PIK	Potsdam Institute for Climate Impact Research
SADC	Southern African Development Community
SAM	Social Accounting Matrix
SDGs	Sustainable Development Goals
SEI	Stockholm Environment Institute
SFC	Specific fuelwood consumption
SPAM	Spatial Production Allocation Model
SOC	Soil organic carbon
tC	Ton carbon
TFP	Total factor productivity
US$	United States Dollar
UN	United Nations
UNEP	United Nations Environment Programme
WEAP	Water Evaluation and Planning

List of tables

Table 2.1: Structure of an aggregated social accounting matrix (SAM) 55
Table 2.2: 2010 Macro SAM for Malawi (millions of kwacha) 66
Table 3.1: Sugarcane-ethanol production technologies 87
Table 3.2: Biofuel and existing crop production technologies, 2010 92
Table 3.3: Production and price impacts 96
Table 3.4: Labor and household impacts 98
Table 3.5: Emissions and water use 102
Table 3.6: Comparing biofuels to alternative cash crops 106
Table 3.A1: Model indices, variables and parameters 112
Table 3.A2: Model equations 113
Table 4.1: Agriculture in Malawi 120
Table 4.2: Estimated means and standard deviations by crop model crop 124
Table 4.3: List of scenarios 128
Table 4.4: Production and price impacts 131
Table 4.5: Labor and household impacts 134
Table 4.6: Investment returns 136
Table 4.A1: Crop yield correlations with maize yields, 1961–2013 147
Table 5.1: Annual per capita fuel demand per region and quintile in 2010 for three stone fire 160
Table 5.2: Demand for fuelwood in Malawi in 1000 metric tons 161
Table 5.3: Sustainable supply of fuelwood in Malawi in 2010 in 1000 metric tons 164
Table 5.4: Demand for fuelwood in 1000 metric tons with increased improved cookstove adoption 167
Table 5.5: Agroforestry system choice as share of maize fields 172
Table 5.6: Fuelwood and maize produced from agroforestry on maize fields in 1000 metric tons 175
Table 5.A1: Specific fuel consumption by stove, fuel and food group 183

Table 5.A2: Linear expenditure system 183
Table 5.A3: Sustainable yield of firewood of different land cover classes in Malawi 185
Table 5.A4: Assumptions for Malawi agroforestry system choice model 186

List of figures

Figure 1.1: The food-energy-water nexus and global drivers 26
Figure 2.1: Circular flow of income in an open economy 43
Figure 2.2: Modeling framework for analyzing biofuel production 71
Figure 2.3: Modeling framework for analyzing irrigation expansion 73
Figure 2.4: Modeling framework for analyzing the biomass energy sector 75
Figure 4.1: Box-whisker-plots showing the GDP gain distribution relative to a non-irrigation baseline 141
Figure 4.2: Box-whisker-plots showing the GDP gain distributions relative to a with irrigation but without fertilizer scenario 142
Figure 5.1: Impact of population and GDP growth on demand and supply of biomass 154

1 Introduction

In November 2011 researchers and policy makers gathered in Bonn, Germany, to discuss the future of food, water, and energy in a globalized world of high population and economic growth, both of which are deemed to put increasing pressure on scarce resources. In the light of these and additional challenges like climate change and urbanization the consensus at the conference in Bonn was that development policy cannot continue on its current "silo" path but must undergo a transformation towards a nexus perspective of integrated water, energy, and food security policies (Hoff, 2011). This study aims at contributing to this transformation through developing innovative modeling approaches for comprehensive and holistic policy analyses. Using Malawi as a case study, the overriding objective is to identify those policy measures that maximize the synergies for food, energy, and water security and minimize the tradeoffs.

1.1 Background: global challenges and the food-energy-water nexus

Human livelihood and survival is dependent on the provision of three fundamental goods: water, food, and energy. The three goods are intrinsically tied together and dependent on each other through consumption and production linkages and through their reliance on limited resources – water, soil and land. For several decades these resources and in turn, food, energy, and water security have come under increasing pressure through growing global challenges, making it crucial to view the three systems from a nexus perspective that captures their interconnections and avoids inefficient resource use arising from single-sector thinking.

Rapid population and economic growth are the predominant drivers of the ever increasing demand for food, water, and energy. Improvements in living standards of a fast growing middle class in many emerging and developing economies as well as urbanization are leading to more resource-intensive consumption patterns, while the urban poor often live in slums without access to safe water, energy, and food (Hoff, 2011). At the same time, the widespread sectoral interconnections through globalization and trade reinforce competition over scarce resources across countries. On the supply side, degradation of the environment through pollution and unsustainable harvesting are threatening an already limited resource base (Wakeford et al., 2015). While these drivers affect demand and supply of food, energy, and water exogenously, climate change poses a more complex challenge through an endogenous two-way relationship (Hoff, 2011). On the one hand energy use and changes in land use through land clearing for food production can lead to greenhouse gas emissions and accelerate climate change. On the other hand, increased climate variability influences agricultural productivity and water supplies, while climate change mitigation measures like the cultivation of biofuels intensify the demand for land and water. These global drivers are of particular importance in fast growing developing and emerging economies, where hundreds of millions of people simultaneously suffer from food, energy, and water insecurity (Ringler et al. 2013). In contrast, population and even gross domestic product (GDP) growth are stagnating in most industrialized countries. As many developed countries are located in the Northern hemisphere, climate change is not yet as prevalent as in the tropics. Developing countries are thus much more affected by global challenges and are additionally facing many local challenges. Economies in developing countries are mostly dominated by the agricultural sector that is heavily dependent on the resources of water and land, but often the same countries face high water loss and land degradation (Penning de Vries et al., 2003). As both food and energy in the form of biofuels are produced by agriculture, the sector is at the heart of the nexus and most affected by resource constraints. The situation is further aggravated as productivity is lower and resources are used less efficiently compared to industrialized countries (Isaksson et al., 2005). Similarly, governance and law enforcement especially over common resources remain a serious issue (Nielsen et al., 2015).

In light of these imminent pressures, an understanding of the interconnections between food, energy, and water systems becomes vital to foster integrated policy making that avoids resource inefficiencies and internalizes externalities. These linkages are manifold and exist along the whole value chain of food, energy, and water systems (Wakeford et al., 2015). For example, water is an essential input in both food and energy production. Inversely, energy is needed to extract water for consumption as well as for irrigation. Food crops can be used for the production of bioenergy, while energy is needed for the production of fertilizer and to convey irrigation water. Chapter 1.3 examines the nexus linkages in detail and identifies potential tradeoffs and synergies that might emerge as externalities from their interdependencies. Tradeoffs arise when the production of one good compromises the secure provision of the other goods through competition for resources. The production of biofuels from agricultural crops for example might crowd out food production from limited land and water and thus threaten food security. At the same time biofuel production could exhibit synergies between energy and food security if farmers benefit from high value biofuel crop production and can afford more and better food.

Nielsen et al. (2015) propose a definition specifically emphasizing the synergies and tradeoffs between the three systems to promote communications between stakeholders and to facilitate policy formulation: "The food-energy-water security nexus encompasses synergies and trade-offs between food, energy, and water security that are impacted by endogenous and exogenous drivers and cannot be captured if these sectors are analyzed in isolation (cf. Nielsen et al., 2015, p. 1)." Other authors such as Wakeford et al. (2015) simply define the nexus as the broad interconnections between food, energy, and water systems originating from production and consumption linkages. Figure 1.1 briefly illustrates the concept of the reciprocal interconnections with synergies and tradeoffs emerging from the use of limited resources and the simultaneous influence of drivers. Both definitions will guide the work of this study, especially as integrated policy making requires an analysis of the tradeoffs and synergies arising from these interconnections at different levels of the economy. Security is multidimensional and needs to be given both at the macro, i.e. national, and the micro, i.e. household or individual level. Food security includes macro-level food security as the availability of food for a country's population through domestic production, stock levels and net imports (FAO, 2008). Micro-level food security means the financial and physical access to sufficient and nutritious food as well as utilization of food depending on nutrient intake and health status. Access to food is thus strongly related to income and poverty. Macro- and micro-level food security are both governed by a fourth dimension of security in the form of stability over time. Similar dimensions apply to energy and water security. The International Energy Agency (IEA) terms energy security as "the uninterrupted availability of energy sources at an affordable price" (cf. IEA, 2013), referring to macro availability, micro access and overall stability of energy. UN-Water specifies water security as "the capacity of a population to safeguard sustainable access to adequate quantities of acceptable quality water [...]" (cf. UN-Water, 2013). The multidimensionality of nexus securities provides an additional challenge for identifying policies that maximize synergies. Without macro-level security, micro-level security and stability is difficult if not impossible to achieve. At the same time ensuring macro-level security does not necessarily translate into access and appropriate utilization. In the worst case a policy aiming to increase macro-level energy security such as the production of biofuels from crops could negatively affect micro-level energy security, for example, if forests are cleared for biofuel production and households lose access to essential firewood for cooking (Nielsen et al., 2015).

Figure 1.1: The food-energy-water nexus and global drivers

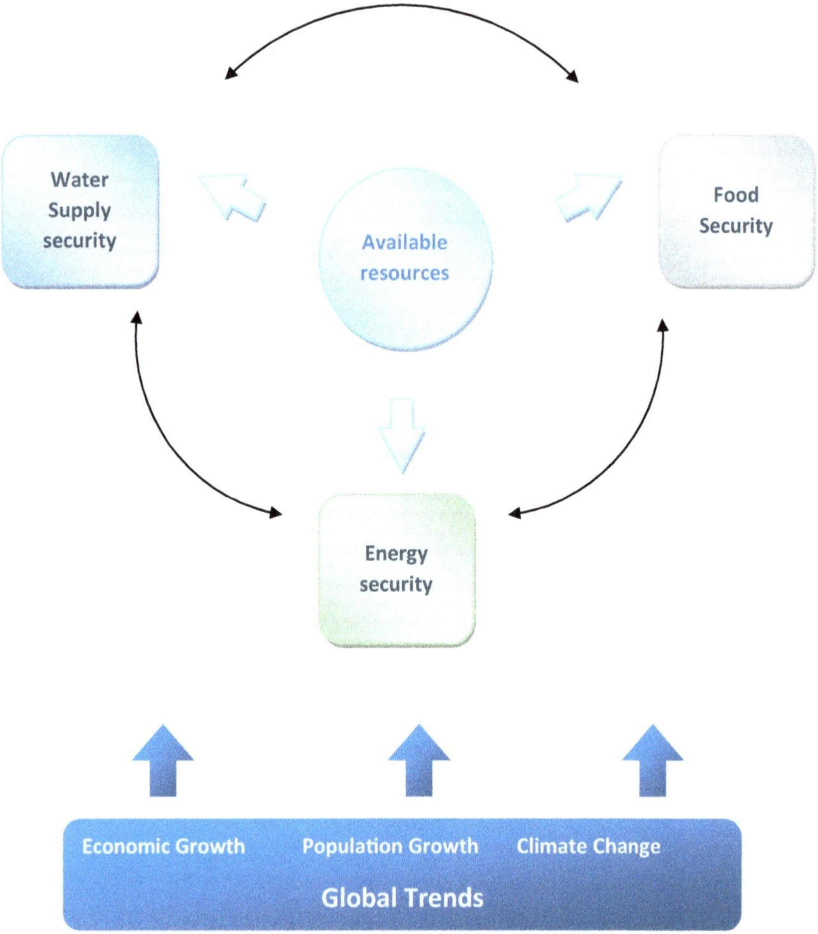

Source: Own creation after Nielsen et al. (2015) and Hoff (2011)

The concept of security not only encompasses different levels of the economy, but also the social and ecological sphere. Availability is related to economic output but also to natural resource availability that can only be achieved as long as the environment is intact. Access and utilization on the other hand depend on the social conditions such as poverty and knowledge and education about the right utilization as well as physical access in terms of market access or access to natural resources. Stability is contingent on social stability as well as environmental factors such as weather variability. In this sense the food-energy-water nexus (FEW nexus) becomes a nexus at different dimensions in space and time, in particular the economy, the environment and social conditions, which are not accidentally the three dimensions of sustainability. These parameters can also be seen as exogenous and endogenous drivers that increase the complexity of the nexus. Drivers thus include the above-mentioned global and local challenges but also parameters like technological innovations, prices or poverty. Most drivers are endogenous, directly affecting the three nexus securities, and are reciprocally affected by them through feedback and feed-forward loops (Nielsen et al., 2015). For example, poverty directly affects and is an inherent part of the access dimension of security, while higher food, energy, and water security usually go in hand with lower poverty. Exogenous drivers especially at the macroeconomic level can also indirectly affect food, energy, and water by creating spillovers through price and market mechanisms. Nielsen et al. (2015) give an example of how exchange rate fluctuations can change the current account balance through impacting imports and exports. Yet, from a macroeconomic perspective, fluctuations of the exchange rate are an endogenous nexus driver as these fluctuations themselves are affected by the level of imports and exports: low food security in terms of availability or domestic production of food could trigger increased imports of food leading the exchange rate to adjust to maintain current account balance. The broader the perspective on the economy and the longer the time horizon, the more drivers become endogenous. At some point, food, energy, and water security could be so low that people die or that economic output stagnates and population and economic growth become endogenous parameters. How wide or narrow the FEW nexus is treated should be dependent on the research question at hand. If policy impacts are only analyzed at the micro-level, macroeconomic parameters remain exogenous. If policies are likely to affect the whole economy, analyses should include more endogenous parameters. This issue is examined in detail in Chapter 2 where a new modelling framework for nexus analysis is developed. As the nexus itself is a part of the broader debate on sustainable development, the discussion of the latter in the next section will help to define the scope of the nexus.

1.2 The FEW nexus in the broader debate on sustainable development

The knowledge that global trends are threatening a finite resource base and that interlinkages between sectors reinforce competition over scarce resources is certainly not new. In a way the FEW nexus can be seen as old wine in new skins and as part of the broader debate on sustainable development. The interconnectedness of the economic, social, and environmental spheres and the potential impacts of resource limits, pollution, and exponential population and economic growth were first demonstrated by the Club of Rome *The Limits to Growth* publication in 1972 (Meadows et al., 1972). The book already contains detailed descriptions on the physical constraints to food production through land and water as well as the inherent tradeoffs between food and the production of other goods and on the competition over limited resources. The authors employed a simple global model considering many of the above-mentioned feedback loops to show that only considerable effort in terms of technological advances and self-imposed limits on economic and population growth would be able to ensure a sustainable world system that is able to satisfy the basic needs of the world population beyond the year 2100 (Meadows et al., 1972). Even though the study warned of the inevitable collapse of the humankind once resources have run out, we are still dealing with the same issues without having found an appropriate solution.[1]

The famous Brundtland report, *Our common future*, brought the themes and findings from *The Limits to Growth* on the global and political stage and shaped the idea of sustainable development as "development that meets the needs of the present without compromising the ability of future generations to meet their own needs" (UN, 1987). The definition emphasizes the needs of the poor in developing countries as well as human-made pressure on scarce resources, since the report sees poverty, inequality and economic growth as the main drivers of environmental degradation.[2] Food, energy, and water are already highlighted as essential needs. Like the Club of Rome, the Brundtland report stresses the

---

1 Nielsen et al. (2015) discuss in detail why it is difficult for policy makers to adopt an integrated policy approach to resource management stemming from both the problem of managing common pool resources as well as from barriers to interministry collaboration.

2 These days it is by no means proven that the so called "downward spiral" of environmental overuse through poor people and subsequent reinforced poverty through environmental poverty really exists (e.g. Dasgupta et al., 2005).

need for demographic change and inclusive growth as a result of holistic policy-making that "[integrates] economic and ecological considerations in decision making" (UN, 1987). Since the publication of the Brundtland report, sustainable development found its way into numerous international agendas and treaties and finally became a principle of the United Nations Millennium Declaration in 2000, manifested in the Millennium Development Goals (MDGs) (UN, 2000). While the MDG of environmental sustainability includes water security and the reduction of biodiversity loss, the main focus of the MDGs was on poverty and hunger eradication. Although a lot of MDG targets were reached on the global level, especially Sub-Saharan Africa stayed behind and still faces high poverty and food insecurity (UN, 2015a). As a consequence, by 2015 the MDGs were succeeded by the Sustainable Development Goals (SDGs) with a clear commitment to sustainable development and even higher aims than the MDGs. The 17 SDGs with their 169 targets are a comprehensive set of goals encompassing the economic, social, and environmental dimensions. Sustainability is thus part of every goal and also explicitly included in the broader goals, such as inclusive education, gender equality, and peace. Food, energy, and water security feature prominently within the seventeen goals, but the SDGs stay true to the Brundtland report and the MDGs in a sense that the first and most important goal is complete poverty eradication until 2030. Even though the SDGs go much further than the MDGs in aiming at eradicating poverty in all its dimensions, some authors have argued that ending hunger should be a top priority for moral and economic reasons (Fan and Polam, 2014).

Indeed, the SDGs have been subject of a lot of critique and "the throw everything into the same pot" approach is likely to make it difficult for policy makers to implement the right policy measures and to set priorities (Easterly, 2015). The FEW nexus approach could compensate for this excessive complexity and allow focused policy-making without compromising any of the SDGs. The International Council for Science (ICSU) and the International Social Science Council (ISSC) discussed the SDGs from a scientific perspective and have criticized the SDGs for lacking an ultimate goal (ICSU & ISSC, 2015). Nevertheless, if – as reasonably assumed – the ultimate goal of the SDGs is sustainable development, which as defined by the Brundtland report means the provision of basic human needs for the poor and the preservation of limited resources, then food, energy, and water security become their natural core. Another important critique of the SDGs is the adherence to "silo" thinking with 17 separate goals and their respective targets that do not consider synergies and tradeoffs with other goals and targets (ICSU & ISSC, 2015). The FEW nexus approach on the other hand was devised to exactly avoid isolated policy-making and to include potential externalities. The nexus can function as a role model how to implement sustainable development policies in an integrated framework while setting the right priorities on fundamental human and environmental needs. Sustainable development as such is not at the center of this work, but in the long run the achievement of universal food, energy, and water security implies that growth is inclusive and development sustainable. Most of the 17 SDGs are at least implicitly connected to the FEW nexus such as the eradication of poverty, responsible consumption and production and climate action. The implementation of policies aimed at simultaneously increasing food, energy, and water security will thus foster the achievement of most of the SDGs. This study seeks to identify and analyze these policy measures in order to find solutions to the inevitability of resource constraints that researchers have been alerting to for more than 40 years. As the impact of policy interventions on food, energy, and water is determined by numerous economic and environmental linkages the next section examines the nexus interconnections in detail.

1.3 The role of linkages in the FEW nexus in detail: economic and ecological linkages

Nielsen et al. (2015) define nexus policies as interventions that directly affect at least one nexus dimension. Such policies are usually aimed at increasing either food, energy, or water security. Through multiple interconnections between the three nexus sectors, those policies invariably affect all three sectors. Most of these linkages are of economic nature in terms of production and consumption connections, resulting in input and/or output linkages through the simultaneous use of natural resources, others originate from ecological processes.

Before examining these interconnections in detail, it is useful to distinguish various country types, as the importance of linkages can differ substantially across nations depending on type of economy and resource endowments. The FAO proposes to assess the FEW nexus following a typology of four different types: two types of agriculture-based economies, either dry or water-rich; industrial countries with natural resource constraints; and transition countries with high population growth and a predominantly industrial economy (Flammini et al., 2014). This typology is somewhat arbitrary in terms of resource endowments and the role of drivers. Especially the distinction between water rich and water scarce agrarian countries is not always useful, as even in countries with large water bodies, the location of lakes or rivers is often not where water is most needed. As resource endowments are very country specific, an overall typology in terms of natural resources is not expedient. Wakeford et al. (2015) suggest two characteristic categories for developing and emerging countries including an agrarian regime with rainfed subsistence farmers relying on traditional biomass energy, and an industrial regime with heavy reliance on fossil fuels for industry and agriculture. The former regime can be found in many countries in Sub-Saharan Africa and will be the focus of this study on the example of Malawi. Due to the subsistence nature in agrarian regimes, the relationship between the environment and the economy is of utmost importance as each single household is directly dependent on and affected by natural resources and climate.

The linkages between the three nexus sectors in agrarian developing countries are less technical and play a stronger role for socio-economic development than in industrial countries. Cooking energy from biomass is a decisive input for clean water as there is basically no formal water distribution system outside of larger cities (Flammini et al., 2014). In more developed countries, energy is needed for lifting, treating and distributing water, such as groundwater pumping, desalination and irrigation (Hoff, 2011). Reciprocally, water is essential for power generation and energy processing. This includes the extraction of fossil fuels as well as substantial water needs in thermal power plants for cooling. While water is used as an input for energy production, pollution of water from fossil fuel mining represents a negative externality. Water is also needed for renewable energy production in both industrial and agrarian economies. Hydroenergy production from dams and in-stream river turbines is an extremely climate change friendly way to produce electricity. Conversely, building of dams and reservoirs can have severe tradeoffs for livelihoods and ecosystems both within the areas that are flooded and further downstream by reducing both water quality and quantity (Flammini et al., 2014). Especially in agrarian developing countries, rivers are lifelines for farmers for irrigation and drinking water. Increased energy security through hydroenergy projects may thus directly affect both water and food security as all three are dependent on one common resource. Energy can also be produced from biofuel crops such as sugarcane that require large amounts of water often in the form of irrigation (De Fraiture et al., 2008).

The latter highlights the controversial link between food and energy. On the one hand, food crops such as maize and sugarcane are used as inputs for energy production, on the other hand crops grown for both food and fuel are competing for the same inputs of land, labor, and water (Flammini et al., 2014). As a measure to avoid this competition for land, agricultural extensification through deforestation could severely reduce micro-level energy security in agrarian economies that rely on biomass for cooking. Cooking energy however is needed for food utilization. At the same time, energy is a crucial input in food production. Mechanized agriculture and irrigation directly require fuel, whereas transportation and processing of food can be very energy intensive as well (Wakeford et al., 2015). While these energy-food linkages are not so important for subsistence agriculture, energy-intensive fertilizer has become a vital input to increase crop yields and food security in many agrarian developing economies.

Finally, the food-water linkage is probably the most important connection as neither crop, vegetable nor livestock production and processing is possible without the input water (Wakeford et al., 2015). In most agrarian developing countries, agriculture is rainfed and low-yielding, while irrigation is restricted to high-value cash crops. Conversely, increased food production can negatively affect water security through both intensification that requires more water input as well as extensification, since deforestation might indirectly decrease water security by inhibiting the ability of the soil to absorb rain water (Hoff, 2011). Water itself is a natural resource whose supply is contingent on the climate system. At the same time, water demand of crops in the form of evapotranspiration depends on climate features such as temperature and humidity as well as soil characteristics (Allen et al., 1998). Ecological processes thus largely determine the role of water in the FEW nexus not least through the interaction between water and nutrients. As such, water can only increase crop yields if soils are fertile (Drechsel et al., 2015). Irrigation might thus require simultaneous fertilizer application to bring benefits for food security, which reinforces the dependence on energy inputs.

These input-output and ecological linkages are complemented by numerous indirect linkages with other sectors that work through markets and are inherent to every economy, leading to a complex set of interconnections.

1.4 Objectives and research questions

As the previous section has shown, the FEW nexus involves linkages and interconnections on different levels and spheres, making it difficult to foresee all potential implications of policy actions. As the complex nexus structure itself determines the effectiveness of policy measures, **the first and methodological objective of this study is to develop an integrated modeling framework for policy analysis that is able to capture the relevant linkages of the food-energy-water nexus**. As will be examined in Chapter 2, existing models that can simultaneously analyze policy impacts on the three nexus sectors are scarce and not able to encompass the large number of impact channels relevant for different policy measures. An integrated and interdisciplinary modeling approach working on different levels of the economy and the environment is still missing from the literature and this study aims at closing this methodological research gap to enable comprehensive and sustainable policy analysis.

A holistic modeling framework is vital to achieve **the second and empirical objective of this study, which is to identify those interventions that maximize synergies and minimize trade-offs between food, energy, and water security**. As will be discussed in the following chapters, previous studies assessing policy impacts on food, energy, and water security are in large part sector-specific. However, a simultaneous analysis of policy impacts on food, energy, and water is essential to identify potential tradeoffs and synergies. The second major contribution of this study to the literature is thus the simultaneous assessment of economic, social, and environmental impacts of policy interventions within the context of food, energy, and water. Four policy measures – biofuel production, irrigation expansion, improved cookstoves and agroforestry – are assessed in detail on the example of Malawi, a developing country in Sub-Sahara Africa. The policy interventions are chosen for three important reasons:

- They directly affect at least two nexus securities thereby providing a large scope for realizing synergies and minimizing tradeoffs.
- Their effectiveness depends on important drivers, in particular resource endowments, population growth, economic growth and climate change.
- They directly impact human livelihoods of subsistence farmers in agrarian developing countries and are thus of the utmost importance for the world's poorest.

The policy analyses are thus guided by three overarching research questions:

1. What is the simultaneous impact of policy measures on food, energy, and water security?
2. What is the role of drivers in the effectiveness of policy measures?
3. How do these policies affect the livelihoods of the poorest?

In addition, the policy measures analyzed comprise particular issues that require more specific research questions, which are examined in the following overview.

1.5 Policy measures analyzed and specific research questions

Bioenergy produced from crops has already been mentioned in the previous sections as an important issue within the FEW nexus. This is because increased biofuel production is a policy that directly affects food, energy, and water security. A positive impact on energy security is straightforward, as domestically produced biofuels should make a country less dependent on fossil fuel imports. In contrast, the biofuel production process is much more water intensive than that of fossil fuels, in particular as biofuel crops such as sugarcane or maize are usually irrigated (King & Webber, 2008). Biofuels are most contested for their competition with food crops for natural resources, especially land and water, and are at least partly blamed for the rise in global food prices in the 2000s (Rosegrant et al., 2008). While these potential tradeoffs weigh heavily, biofuels have been found to have positive effects on rural development, poverty and income, thereby indirectly increasing the access dimension of food security (e.g. Arndt et al., 2012; Negash & Swinnen, 2013). Although land clearing and land use change for biofuel crops might increase greenhouse gas (GHG) emissions (Fargione et al., 2008), some biofuel crops such as sugarcane are carbon sinks and could contribute to climate change mitigation by increasing soil carbon sequestration, leading to much lower emissions per liter compared to fossil fuels. Overall, evidence on the negative impacts of biofuel production remains ambiguous. Zilberman et al. (2013) for example find no clear impact from global biofuel production on national food prices. While water-scarce countries should not engage in water-intensive bioenergy production, several studies find that global water resources are still sufficient for an expansion of biofuels (e.g Berndes 2002; De Fraiture et al., 2008). As most of these studies are sector-specific focusing either on economic or environmental outcomes or food security, they omit linkages between food, water, energy and potential drivers, and thus cannot capture important synergies and tradeoffs.

Against this ambivalent background, biofuel crops experience much larger scrutiny than other export crops by policy makers, in particular the European Union (EU). The EU's sustainability criteria specify that biofuels imported into the EU have to generate at least 40 percent fewer GHG emissions than fossil fuels per liter and have to minimize displacement of food crops and exorbitant water use (EC, 2010). This higher scrutiny is problematic for developing countries as the EU is an important export market due to preferential trade agreements for low income countries such as Malawi. In light of the mixed evidence concerning tradeoffs from sector-specific studies and the potential benefits of biofuel production for developing countries, several research questions arise apart from the general research questions guiding the policy analyses in this study. In particular, does biofuel production necessarily displace food crops or could a positive impact on income and economic development increase food security? What is the impact of biofuels on GHG emissions and climate change mitigation and are biofuels produced in developing countries able to meet the EU's sustainability criteria? Do biofuels deserve the scrutiny from policy-makers in a sense that they are worse than other export crops for food security and the environment? How is the water intensity of agriculture affected by biofuel production?

The last research question is of particular importance as a lot of high value crops are irrigated. Irrigation expansion is an essential policy measure for raising crop yields and food security, especially as a climate change mitigation mechanism to contain the impacts of weather variability (Svendsen et al., 2009). As the latter is likely to have negative effects on water security, the potential for irrigation expansion is dependent on a country's water resource endowments. Increased irrigation could indirectly lead to lower energy security through decreasing water availability for hydroenergy production. Conversely, large-scale irrigation techniques require a lot of energy for pumping and distributing water. This tradeoff can be alleviated using mainly smallholder irrigation techniques such as gravity irrigation or treadle pumps that are more affordable to poor farmers and have a higher potential to increase food security (SMEC, 2015). Even though enough water would be available in many countries in Sub-Sahara Africa, irrigation use still remains extremely low (You et al., 2011). The investment in irrigation expansion hinges on its profitability, and a lot of irrigation projects in Sub-Saharan Africa did not manage to cover the investment costs due to lower than expected yields and high labor requirements (Inocencio et al., 2007). Previous studies assessing the returns to irrigation do not consider the simultaneous value of direct impacts on income and output on the one hand and indirect impacts such as multiplier effects arising from sectoral interconnections and drivers as well as agro-ecological linkages and climate risk reduction on the other hand (e.g. Gebregziabher et al., 2009; Dillon, 2011; You et al., 2011). Thus in addition to the guiding research questions, the second empirical study in this book simultaneously assesses the benefits of irrigation expansion originating from all potential impact channels and linkages to evaluate whether irrigation investments can be profitable. In particular, can irrigation be profitable when producing food crops for higher food security or would irrigation expansion lead to food crop displacement in favor of higher value crops? In terms of climate change mitigation, what is the potential of irrigation to reduce risk and vulnerability of poor farmers? What is the importance of agro-ecological conditions and crop management techniques for increasing crop yields?

Importantly, an increase in food security through irrigation will go hand in hand with increased demand for energy for cooking. Domestic energy in Sub-Saharan Africa is not so much a question of modern fossil or renewable energy, but largely dependent on natural resource endowments in terms of wood. Biomass energy in the form of firewood and charcoal is still the dominant source of cooking energy in Sub-Saharan Africa and is thus vital for food security (IEA, 2014). Meanwhile, population and GDP growth are putting increasing pressure on a dwindling supply base, both through increasing demand for fuelwood and through deforestation for agricultural expansion. Two policies are propagated to alleviate the supply-demand imbalances. On the demand side, the distribution of improved cookstoves aims at higher energy efficiency of cooking appliances and subsequently lower demand for biomass energy. This could directly increase energy security as people would have lower fuel needs and increase food security through better cooked food (Lewis & Pattanayak, 2012). On the supply side, agroforestry allows biomass energy production alongside crops on farmers' fields and directly increases the resource base. The latter is a win-win policy measure for all nexus securities: Agroforestry would directly increase energy security as people produce firewood themselves and do not have to collect wood from forests. Most agroforestry practices involve fertilizer trees that have positive effects on crop yields and thus food security (Garrity et al., 2010). In addition, trees and forests are essential for watershed protection and carbon sequestration, thereby increasing water security and lowering GHG emissions.

Despite these promising impacts, policy-makers in Sub-Saharan Africa and worldwide largely ignore the importance of traditional biomass energy. The SDGs for example promote only modern and renewable energy as a means to sustainable development (UN, 2015b). Biomass energy on the other hand can be inherently renewable if produced sustainably. In addition to assessing the general research questions, the aim of the third empirical study is to show the importance of biomass energy in terms of sustainable development, especially considering the essential linkages of biomass energy with food security and the environment. In particular, the study assesses the potential of improved cookstoves and agroforestry to establish a sustainable biomass energy sector in the light of population and economic growth.

The effectiveness of all four policy measures analyzed in this study is strongly dependent on the linkages between food, energy and water as well as with other sectors and the environment. The findings of the empirical studies are only valid if all relevant impact channels are included in the analysis. Using the right framework for policy assessment is thus crucial for capturing tradeoffs and synergies.

1.6 Structure of the study

The study proceeds with the following outline. The next chapter reviews existing modeling approaches and explains the methodologies developed for analyzing the four policy measures examined above. This is followed by the application of the innovative modeling frameworks in three empirical studies that assess the four policy interventions in detail and answer the research questions posed in the previous section. Chapter 3 explores the impact of increased biofuel production from sugarcane compared to other export crops on the FEW nexus in Malawi by simultaneously analyzing economic and environmental impacts. This is followed in Chapter 4 by evaluating the benefits of irrigation investments under uncertainty. Chapter 5 examines the biomass energy sector in Malawi and analyzes the potential of improved cookstoves and agroforestry for simultaneously increasing food and energy security. Chapter 6 summarizes the methodological contributions and empirical findings of this study. In addition, more general policy implications are derived and directions for future research identified.

References

Arndt, C., Pauw, K., Thurlow, J., 2012. Biofuels and economic development: A computable general equilibrium analysis for Tanzania. *Energy Economics*, 34(6), 1922–1930.

Berndes, G., 2002. Bioenergy and water: The implications of large-scale bioenergy production for water use and supply. *Global Environmental Change*, 12, 253–271.

Dasgupta, S., Deichmann, U., Meisner, C., Wheeler, D., 2005. Where is the Poverty – Environment Nexus? Evidence from Cambodia, Lao PDR, and Vietnam. *World Development*, 33 (4), 617–638.

De Fraiture, C., Giordano, M., Liao, Y., 2008. Biofuels and implications for agricultural water use: blue impacts of green energy. *Water Policy*, 10(S1), 67–81.

Dillon, A., 2011. The Effect of Irrigation on Poverty Reduction, Asset Accumulation, and Informal Insurance: Evidence from Northern Mali. *World Development*, 39 (12), 2165–2175.

Drechsel, P., Heffer, P., Magen, H., Mikkelsen, R., Wichelns, D. (Eds.) 2015. Managing Water and Fertilizer for Sustainable Agricultural Intensification. International Fertilizer Industry Association (IFA), International Water Management Institute (IWMI), International Plant Nutrition Institute (IPNI), and International Potash Institute (IPI). First edition, Paris, France.

Easterly, W., 2015. The SDGs Should Stand for Senseless, Dreamy, Garbled. Foreign Policy, September 28, 2015. http://foreignpolicy.com/2015/09/28/the-sdgs-are-utopian-and-worthless-mdgs-development-rise-of-the-rest/, access date: 30.07.2016.

EC (European Commission), 2010. Communication from the Commission on the practical implementation of the EU biofuels and bioliquids sustainability scheme and on counting rules for biofuels. *Official Journal of the European Union*, C160, 8–16.

Fan, S., Polman, P., 2014. An ambitious development goal: Ending hunger and undernutrition by 2025. In: 2013 Global food policy report. Eds. Marble, A. and Fritschel, H., Chapter 2. Pp. 15–28. Washington, D.C.: International Food Policy Research Institute (IFPRI).

FAO, 2008. An Introduction to the Basic Concepts of Food Security. Food Security Information for Action, Practical Guides. Published by the EC - FAO Food Security Programme, Rome, Italy.

Fargione, F., Hill, J., Tilman, D., Polasky, S., Hawthorne, P., 2008. Land clearing and the biofuel carbon debt. *Science,* 319, 1235–1238.

Flammini, A., Puri, M., Pluschke, L., Dubois, O., 2014. Walking the Nexus Talk: Assessing the Water-Energy-Food Nexus in the Context of the Sustainable Energy for All Initiative. Environment and Natural Resources Working Paper No. 58, FAO, Rome, Italy.

Garrity, D., F. Akinnifesi, O. Ajayi, S. Weldesemayat, J. Mowo, A. Kalinganire, M. Larwanou, and J. Bayala, 2010. Evergreen Agriculture: a robust approach to sustainable food security in Africa. *Food Security*, 2 (3), 197–214.

Gebregziabher, G., Namara, R.E., Holden, S., 2009. Poverty reduction with irrigation investment: An empirical case study from Tigray, Ethiopia. *Agricultural Water Management*, 96, 1837–1843.

Hoff, H. 2011. Understanding the Nexus. Background Paper for the Bonn 2011 Conference: The Water, Energy and Food Security Nexus. Stockholm, Sweden: Stockholm Environment Institute.

ICSU, ISSC, 2015. Review of the Sustainable Development Goals: The Science Perspective. Paris: International Council for Science (ICSU). Paris, France.

IEA (International Energy Agency), 2013. Topic: Energy Security. http://www.iea.org/topics/energysecurity/, access date: 10-09-2013.

IEA (International Energy Agency), 2014. Africa Energy Outlook, A Focus on Energy Prospects in Sub-Saharan Africa, World Energy Outlook Special Report. Paris, France.

Inocencio, A.; Kikuchi, M.; Tonosaki, M.; Maruyama, A.; Merrey, D.; Sally, H.; de Jong, I. 2007. Costs and performance of irrigation projects: A comparison of sub-Saharan Africa and other developing regions. Colombo, Sri Lanka: International Water Management Institute. 81 pp. (IWMI Research Report 109).

Isaksson, A., Ng Hee, T., Robyn, G., 2005. Productivity in Developing Countries: Trends and Policies. UNIDO Research Programme, United Nations Industrial Development Organization, Vienna, Austria.

King, C., Webber, M., 2008. Water Intensity of Transportation, *Environmental Science & Technology*, 42 (21), 7866–7872.

Lewis, J.J., Pattanayak, S.K., 2012. Who adopts improved fuels and cookstoves? A systematic review. *Environmental Health Perspectives*, 120 (5), 637–645.

Meadows, D. H., Meadows, D. L., Randers, J., Behrens III, W. W., 1972. The Limits to Growth: a report for the Club of Rome's project on the predicament of mankind. Potomac Associates, Washington DC, USA.

Negash, M., Swinnen, J.F.M., 2013. Biofuels and food security: Micro-evidence from Ethiopia. *Energy Policy*, 61, 963–976.

Nielsen, T., Schuenemann, F., McNulty, E., Zeller, M., Nkonya, E., Kato, E., Meyer, S., Anderson, W., Zhu, T., Queface, A., Mapemba, L., 2015. The Food-Energy-Water Security Nexus: Definitions, Policies, and Methods in an Application to Malawi and Mozambique. Discussion Paper 01480, Washington, DC: International Food Policy Research Institute.

Penning de Vries, F., Acquay, H., Molden, D., Scherr, S., Valentin, C., Cofie, O., 2003. Integrated land and water management for food and environmental security. Comprehensive Assessment of Water Management in Agriculture Research Report 1. Colombo, Sri Lanka: Comprehensive Assessment Secretariat.

Ringler, C., A. Bhaduri, and R. Lawford. 2013. The Nexus across Water, Energy, Land and Food (WELF): Potential for Improved Resource Use Efficiency? *Current Opinion in Environmental Sustainability*, 5, 617–624.

Rosegrant, M.W., Zhu, T., Msangi, S., Sulser, T., 2008. Global scenarios for biofuels: Impacts and implications. *Review of Agricultural Economics*, 30(3), 495–505.

SMEC, 2015. National Irrigation Master Plan and Investment Framework – Main Report for Republic of Malawi, Ministry of Water Development and Irrigation, Department of Irrigation. Lilongwe, Malawi.

Svendsen, M., Ewing, M., Msangi, S., 2009. Measuring Irrigation Performance in Africa. Discussion Paper 00894, Washington, DC: International Food Policy Research Institute.

UN-Water, 2013. *Water Security & the Global Water Agenda*. http://www.unwater.org/downloads/watersecurity_analyticalbrief.pdf, access date: 10-09-2013.

UN, 1987. Report of the World Commission on Environment and Development: Our Common Future (Brundtland Report). United Nations, New York, USA.

UN, 2000. General Assembly resolution A/55/L.2, United Nations Millennium Declaration. United Nations, New York, USA.

UN, 2015a. The Millennium Development Goals Report 2015. United Nations, New York, USA.

UN, 2015b. General Assembly resolution A/70/L.1, Transforming our world: the 2030 Agenda for Sustainable Development. United Nations, New York, USA.

Wakeford, J., Kelly, C. and Mentz Lagrange, S. 2015. Mitigating risks and vulnerabilities in the energy-food-water nexus in developing countries: Summary for Policymakers. Sustainability Institute, South Africa.

You, L., Ringler, C., Wood-Sichra, U., Robertson, R., Wood, S., Zhu, T., Nelson, G., Guo, Z., Sun, Y., 2011. What is the irrigation potential for Africa? A combined biophysical and socioeconomic approach. *Food Policy*, 36(6), 770–782.

Zilberman, D., Hochman, G., Rajagopal, D., Sexton, S., Timilsina, G., 2013. The impact of biofuels on commodity food prices: Assessment of findings. *American Journal of Agricultural Economics*, 95(2), 275–281.

2 Developing the appropriate modeling framework for quantitative policy analysis of the food-energy-water nexus

As the previous chapter has shown, the FEW nexus comprises a complex system of interconnections that determine how policy measures affect food, energy, and water security. The sheer complexity of potential effects makes a quantitative analysis of policy impacts indispensable to evaluate and compare relevant synergies and tradeoffs. Quantitative policy assessments are divided into ex-ante and ex-post analysis. The latter comprises impact evaluations to analyze the impacts of an already implemented policy on specific outcomes such as income or food security. An ex-ante policy analysis uses numerical models that allow for the simulation and comparison of a multitude of policy measures on different levels of the economy. While simulation models can never predict the definite impact of a policy measure, they are crucial to guide policy-makers on potential benefits and tradeoffs and the processes underlying policy impacts (Piermartini and Teh, 2005). Simulation models are therefore ideal tools to identify those policy measures that exhibit the largest synergies between food, energy, and water security. The challenge in modeling the FEW nexus lies in capturing the relevant linkages both of economic, social, and environmental nature. A model encompassing all economic and ecological processes pertaining food, energy and water is – apart from being unmanageable without experts from different disciplines[1] – not expedient, since different policy measures operate through different linkages. Scale and context of the policy measure to be analyzed need to be considered when choosing the right modeling framework. The focus of this research is on national policies in a developing country that most certainly yield economy-wide impacts. In addition, an appropriate modeling framework requires clear system boundaries both in spatial and temporal terms to conduct useful policy analysis (Bazilian et al., 2011). There are numerous modeling approaches that model one or more sectors of the nexus, which will be reviewed in the next sections considering their applicability for analyzing policy impacts on the FEW nexus in Malawi.

The usefulness of models depends on how well they capture the relevant linkages for the policy interventions analyzed. The interconnections between food,

1 As will be examined in a later section, there are several approaches to systems analysis that try to capture all economic and ecological linkages on a global level.

energy, and water are predominantly defined by the competition for limited resources resulting from input and output linkages. These linkages affect all actors and markets within in an economy through the circular flow of goods and services. The structure of an economy thus determines the impact of policy measures, so that the search for a suitable modeling framework needs to begin with understanding the nature of economic linkages.

2.1 The importance of economic linkages

The linkages between economic actors such as households, producers, and the government define the structure of an economy. These interconnections are a result of the circular flow of goods and services, mediated through product and factor markets and influenced by the behavior of economic actors (Dervis et al., 1982). In a non-barter economy, economic linkages correspond to the flow of income in monetary units in exchange for goods and services. Figure 2.1 depicts the flow of income in an open economy with aggregated economic actors represented by rectangles and markets by circles. At the center of this circular flow are market interactions between households or consumers and firms or producers. Households offer their factors of production such as land and labor to producers at factor markets and in turn receive income, which they use to buy goods and services leading to consumption linkages between producers and households. Mediated through product markets, producers receive the households' consumption spending in the form of revenue, which they use to pay for wages and rents at factor markets or for inputs at product markets. As producers and households come together at factor and product markets, the market mechanism balances demand and supply by adjusting prices. Production linkages between different sectors[2]/producers encompass backward and forward linkages. Backward or upstream linkages emerge from intermediate input use of producers, e.g. fertilizer use in agricultural production. If the output of the same producer is used as an input by other producers, a forward linkage between the two producers arises, such as in downstream processing of food that uses agricultural products (compare Diao and Thurlow, 2012). Producers not only trade each other's outputs, but also compete for products and factors at both markets. If that involves scarce and expensive resources, potential tradeoffs and crowding out of producers can arise.

2 Here a sector denotes the sum of producers or firms producing the same commodities such as the agricultural sector, which encompasses many producers that all produce agricultural commodities.

Figure 2.1: Circular flow of income in an open economy

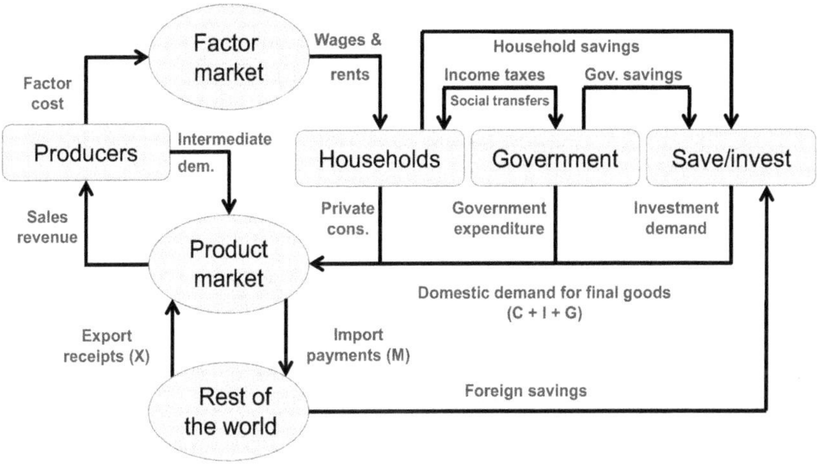

Source: Author's creation after Breisinger et al. (2009)

Resources that are important for food, energy, and water security are both traded at factor markets in terms of land and at product markets such as fertilizer for food production. At the same time, the nexus also includes processes outside of market interactions. Some resources such as rain water originate from the ecosystem and are not directly affected by economic actions. In subsistence economies in rural regions of developing countries, a lot of products such as home-produced food or collected firewood are acquired outside of actual markets. Both processes need to be kept in mind when choosing the appropriate model.

While microeconomic behavior of households and producers dominates market interactions in an economy, macroeconomic institutions and actors are involved in the circular flow of income as well. The government receives taxes from consumers and producers (not shown here), which are used for government consumption, savings and social transfers. Savings serve as capital that is reinvested into production. In an open economy, the rest of the world takes part in market interactions through exports and imports and provides foreign savings. Considering the multitude of linkages originating from the behavior of economic agents and market interactions that are apparent from the stylized economy in Figure 2.1, a policy measure – even if it is aimed at increasing only the output of one sector – will inevitably affect the rest of the economy. The social sphere is thus directly affected by the structure of the economy as policy impacts work through consumption and production linkages.

The importance of economic linkages can best be demonstrated when considering a policy example. In most developing countries of Sub-Saharan Africa, agriculture and the rural sector dominate the economy and exhibit strong linkages to all factor and product markets (Sadoulet and De Janvry, 1995). If biofuel crop expansion aimed at higher energy security is conducted without land clearing, the amount of land available to all other crops will directly decrease. Through a larger demand for land than supply, competition on land markets increases, leading to an imbalance that is solved through the price mechanism. In the end agricultural land becomes more expensive and producers that cannot make a profit under higher land costs have to drop out of the market. If these producers have produced food crops, the output of food crops decreases. These direct impacts on the agricultural sector lead to a multiplier effect that is the sum of indirect impacts on the rest of the economy stemming from both consumption and production linkages (Breisinger et al., 2009). Lower supply of food first meets a stable demand for food at food markets, where the price mechanism leads to higher food prices to reach a new equilibrium of demand and supply. For some households, the new food prices are too high to buy as much food as before, leading to lower demand of food and to lower food security. These households may also decrease their non-food consumption in order to maintain food security, which will negatively affect producers of non-food goods. Other households might benefit from increased biofuel crop production, as they can rent their land for a higher price which leads to increases in income. With higher income they can afford the increase in food prices and demand even more food and non-food goods than before. Consumption linkages can thus trigger structural change in the producing sectors similar to production linkages: If food crop production decreases, downstream processors of food crops lose their jobs and subsequently income. These producers might go into non-food production or biofuel processing, leading to a growth of output in these sectors. In addition, increased demand for inputs from the growing industries stimulates upstream production. As demand for labor in the growing industries is high, wages increase leading to higher income of households employed which in return triggers consumption linkages in a second round. On top of these microeconomic impacts, macroeconomic interactions influence markets and economic actors. The higher consumption expenditure for food may decrease savings. This in turn decreases capital needed for investment, which negatively affects producers. Increased biofuel exports lead to an imbalance in the current account, which is cleared by the exchange rate. The latter determines the competitiveness of exporting producers. Those that are not competitive at the world market after a depreciation or appreciation of the exchange rate will have to cease production.

Policy impacts through economic linkages are thus manifold and may affect every actor in an economy through consumption and production linkages. Environmental linkages such as interactions between soil and climate that affect crop yields, though important for production and output, will always lead to subsequent economic interactions that determine the distributional impacts of policy measures. The changes in the distribution of incomes largely determine policy impacts on the social sphere. As the access of households to food, energy, and water is central to livelihoods and survival, a model capturing relevant economic linkages needs to be at the heart of policy analysis of the FEW nexus. The next section will thus review the usefulness of existing economic simulation models for analyzing the FEW nexus in Malawi.

2.2 Economic policy simulation models and their applicability to analyze the nexus

Economic models replicate the structure of the economy and emulate the behavior of economic agents. A subset of economic models uses optimization algorithms with the goal of finding the optimal allocation of scarce resources under different policy scenarios. These are most suitable for policy simulations in the FEW nexus and are discussed in the following section. Models are built at different levels of the economy, either economy-wide, at the sector level or even at the micro-level. To depict the behavior of economic actors and the structure of the economy, most numerical simulation models have two distinct features: first, like analytical models, they use mathematical equations based on economic theory to emulate behavioral responses. Behavior of rational producers and households is usually governed by the maximization of benefits in the form of utility and profit. Secondly, the parameters of these equations are calibrated with empirical real-world data[3] to mirror the specific economic structure of a country or sector with production and consumption linkages. The size of a model in terms of economic linkages covered depends on the part of the economy studied. The larger the model, the less detailed the microeconomic structure of individual economic agents. Three general model categories can be distinguished for quantitative policy analysis at country level in developing countries: household, partial, and general equilibrium models. The latter are economy-wide models

3 Some parameters need to be estimated from empirical data or taken from existing literature.

that reproduce the structure of an economy by explicitly capturing all economic linkages and respecting both micro- and macroeconomic constraints including income and resources (Piermartini and Teh, 2005). Economy-wide models function at a relatively high level of aggregation, so that single economic actors are usually aggregated to representative agents that are calibrated to the behavior of larger groups of households or producers. General models invariably lose some detail that is better captured by partial equilibrium models. In these models, a part of the economy is depicted in more detail, such as one or more markets for specific commodities. To capture behavioral responses of individuals, farm household models are valuable tools as examined in the following section.

2.2.1 Farm household models

As already mentioned, the agricultural sector dominates many developing economies in Sub-Saharan Africa. In countries like Malawi subsistence agriculture accounts for the majority of households in rural areas. The distributional impacts of policy measures therefore depend largely on the behavioral interactions of semi-commercialized farmers (Singh et al., 1986). While the previous section on economic linkages distinguishes between producers and households/consumers, the distinctive feature of farm households, especially in developing countries, is that they incorporate both types of economic actors. Farm households are usually not completely self-sufficient, so that some food still needs to be bought at the market and some of the family labor available is supplied to labor markets. Conversely, larger farms sell part of their produce at the market and hire additional labor (Taylor and Adelman, 2003). Off-farm labor income is important if households are net buyers of food: especially in Malawi, a lot of farm households do not produce enough food to satisfy their consumption needs due to extremely small farm sizes and labor shortages at the height of the rainy season (Alwang and Siegel, 1999). A policy that is for example aimed at increasing the output of food crops and food availability might thus not lead to higher food security for all. The first round income effect through lower prices is only positive for net buying households, but negative for net sellers of food.

The different on-farm and off-farm production and labor activities as well as the demand structure of farm households are crucial for the success of policy measures in the rural economy. Models that capture these features in detail can therefore provide insight on the appropriate design of pro-poor policies, even if markets are only partly involved in farm household production and consumption decisions. Farm household models (FHM) are truly microeconomic but models exist at different levels of the economy, either single household, village, regional or national. Irrespective of the level, the models center on the integrated production and consumption behavior (including labor allocation) of farm households and capture household behavior with interdependent mathematical optimization problems in two different ways depending on the assumptions. If all factor and product markets are complete, households are price takers in both markets. In this case, separability of production and consumption decisions holds and the model is solved recursively or sequentially with households' utility hinging on production profits (Singh et al., 1986). The perfect market assumption implies that all products and factors are tradable and the opportunity costs of goods produced and factors used correspond to actual market prices and wages. This assumption is difficult to hold due to numerous inherent market failures in the rural economy: Family labor originating from women and children is no perfect substitute to hired labor and may not even be tradable at the labor market. Or markets for certain commodities might simply be not available due to high transaction costs. In these cases, factors and commodities cannot be treated as tradables anymore and their prices do not correspond to market prices, but are determined within the household corresponding to individual shadow prices (Sadoulet and De Janvry, 1995). This is for example the case with fuelwood in rural areas that is collected by women and children. Even if there is a market for fuelwood in cities, the farm household's internal shadow price cannot be equated with the urban market price due to transaction costs and the individual household shadow value of labor of women and children. Separability does not hold in these cases and production and consumption decisions must be modeled simultaneously. In the extreme case that there are no markets at all, farm household models can be seen as very small general equilibrium models where the household's internal shadow price mechanism leads to equilibrium at the household's factor and product markets (Taylor and Adelman, 2003).

The application of FHM to the FEW nexus has been limited so far, although the models can play an important role for analyzing policies on the micro-level, especially in terms of access. As the impact of interventions on agricultural production and income of farm households is the essence of most models, the access and availability dimensions of food security can be easily assessed. Even so, only few FHM explicitly analyze the impact of policies on household food security (Van Wijk et al., 2014). Firewood collection is part of the farm household's production activities making FHM also suitable to assess policy effects on energy security. Until now farm households' energy demand behavior has been mostly studied through ex-post regression analysis (e.g. Pattanayak et al., 2004;

Mekonnen et al., 2015). FHM cannot capture the environmental linkages that affect water security. As will be explained in a later section, water impacts of policy measures need to be modeled with biophysical models.

Farm household models can only give a partial view of economic linkages limited to the rural economy and a certain group of households. Yet, they can be relatively easily built and are essential to analyze policy impacts in the case of market failures as well as specific distributional impacts as they provide higher detail on intra-household interactions compared to aggregated models. Different members of the farm household face different opportunity costs that are likely to play a role for policy impacts (Sadoulet and De Janvry, 1995). Off-farm labor opportunities for men might be higher than for women, while women face the additional burden of reproductive labor. The latter includes domestic activities that usually fall to women such as caring for and rearing children, cooking, maintenance of the household and cleaning (Beneria, 1979). In addition women are usually employed in food crop production, while men tend to cash crops. Policies that subsidize food or cash production might thus lead to unintended intra-household dynamics, where additional food production might even negatively affect other dimensions of food security. Food utilization of children depends crucially on the mother's time for reproductive work leading to a tradeoff between work at the field and at the house. FHM do not encompass all production and consumption linkages in an economy and can thus not capture indirect effects of policy measures arising through different markets. Yet, they are able to assess indirect effects stemming from specific household constraints that would be lost in aggregated models. Moreover, if certain production and consumption decisions are made outside of markets, economy-wide linkages play only a minor role for the impact of policy measures. Depending on the policy analyzed, a partial view of the economy might thus be better suited to identify the most beneficial intervention.

2.2.2 Partial equilibrium models

In general, partial equilibrium models examine a part of the circular flow of income, which can be a whole sector or a single market, while disregarding other markets, their interactions and resource constraints. The main advantages of partial equilibrium models are their reduced complexity through more restrictive system boundaries, the ability to go into meticulous detail concerning the sector they examine, and their ability to provide a good measure of direct effects of policy measures (Sadoulet and De Janvry, 1995). This detail comes at a cost, since partial equilibrium models do not encompass all consumption and production linkages between economic actors. These models are therefore most suitable for quantitative policy analysis, if a policy measure is unlikely to have significant indirect economic effects because of intersectoral linkages (Piermartini and Teh, 2005). Although indirect economic impacts are essential for the FEW nexus, partial equilibrium models can be useful to assess specific parts of the nexus and policy effects in more detail, given their flexibility especially in spatial terms.

There is no standard model, but models differ depending on the problem they study, ranging from single market models to spatial or multimarket models (Piermartini and Teh, 2005). Most models focus on market interactions between producers and consumers. In the simple single-market model, the impact of policy measures is limited to affecting demand and supply through price changes of the respective good in that market. This means that the only endogenous variables of the model are price and quantity of the commodity studied while the rest of the economy is ceteris paribus. The single market price changes do not generate any spillovers for household income and the demand on other factor or product markets. Prices in other markets are treated as constant, while changes in resource allocations and resource constraints are not considered (Piermartini & Teh, 2005). Multimarket or multi-commodity models go a step further and incorporate interactions between a group of goods or factors that exhibit strong intersectoral linkages. They focus on assessing indirect impacts of policy measures on supply and demand of substitutes markets, but can also trace how price and quantity changes affect household incomes and resource allocations (Arulpragasam and Conway, 2003). Spatial multi-market models include trade across regions. Multi-market models can thus capture a large number of economic linkages, but concentrate on market interactions of the specific sectors analyzed. They can thus not capture the whole multiplier effect of policy measures and may underestimate second-round effects, especially as savings and subsequent investments are not considered (Sadoulet and De Janvry, 1995).

Due to their incomplete view of economic linkages, partial equilibrium models are not useful for an integrated analysis of the FEW nexus, but are valuable tools to better understand policy impacts on each specific sector in more detail. Most existing energy market models are aimed at assessing policies for climate change mitigation and thus allow for an integration of both economic and environmental linkages, but do not capture the social sphere and distributional impacts of policies. The MARKAL model of the International Energy Agency (IEA) is a partial equilibrium model of the energy sector that has been mostly applied to industrial countries and can be used at a national or regional scale. Different energy policies and climate change scenarios are simulated to determine which technologies minimize the cost of the energy system and GHG emissions (IEA-ETSAP, 2014). The MESSAGE model is an engineering model that is used to conduct energy policy analysis at a global level and can be integrated with partial equilibrium models for multi-market analysis (IIASA, 2014). An energy model that has been extensively applied to developing countries is the Long-range Energy Alternatives Planning (LEAP) model of the Stockholm Environment Institute (SEI) (SEI, 2013). The model has been used to analyze climate change mitigation policies as well as for forecasting energy demand such as for Tanzania and even Malawi (UNEP, 1999; Njewa, 2012). Nevertheless, the LEAP model is an accounting framework that does not include actual behavior and optimization of economic agents and is therefore no partial equilibrium model per se (Bhattacharyya and Timilsina, 2010). Similarly, the water evaluation and planning (WEAP) model of SEI does not include economic behavior and is mainly a tool for water system planning (WEAP21, 2014). Since water is a natural resource mostly affected by environmental linkages and freely available through rain and groundwater, it cannot be captured by economic models. This is a general problem of water security analysis in the nexus as will be examined in more detail in a later section. WEAP and other water simulation models integrate hydrological models by considering rainfall, streamflow, runoff or evapotranspiration. These features are decisive to capture actual supply and demand of water outside of economic markets.

Policy impacts on food security and food markets can be analyzed with agricultural sector partial equilibrium models that cover markets for all types of agricultural crops as well as production factors especially in terms of land. In this sense, they are multi-market and multi-commodity models with emphasis on different issues such as price projections or policy simulation. Trade plays an important role in agricultural sector models, so that most models run at the regional and global level. Many models focus on a certain region, such as ESIM and CAPRI on agricultural policies in the EU, and study developing countries only at high levels of aggregation (Blanco-Fonseca, 2010). The International Model for Policy Analysis of Agricultural Commodities and Trade (IMPACT) of the International Food Policy Research Institute (IFPRI) on the other hand has a specific focus on food security and poverty in developing countries and can assess policy impacts on macro-level food security on a global, regional, and national scale. The latest version of IMPACT is a holistic modeling system that integrates the original agricultural multi-market model with biophysical models (climate, crop, and water models) (Robinson et al., 2015). The modeling framework thus encompasses both economic and environmental linkages pertaining to agricultural production and covers both water and food security. Regardless, as a partial equilibrium model, IMPACT misses indirect policy effects originating from production and consumption linkages outside the agricultural sector that are essential for distributional impacts especially of non-farm households. The following type of economic simulation models is specifically designed to encompass the complete flow of income in an economy and is thus ideal to examine policy impacts on Malawi.

2.2.3 Computable general equilibrium models

Computable general equilibrium (CGE) models are economy-wide simulation models that are able to assess both direct and indirect multiplier impacts of policy measures. They can analyze policy effects on output, resource allocation and income, and can thus capture both the access and availability dimension of security. CGE models exist on different levels of the economy and can be regional, national, global and even village level. The national level CGE models encompass the complete circular flow of income in an economy and thus all economic linkages and market interactions outlined in Figure 2.1. The special feature of CGE models is their integration of the microeconomic and macroeconomic sphere. Traditionally, there is a clear division between micro- and macroeconomics: microeconomics studies the interaction of individual consumers and producers on markets. Macroeconomics studies the behavior of the economy top-down through aggregates and institutions like GDP, the household sector, international trade, employment and the government (Felderer and Homburg, 2005). In the real economy, macro aggregates are determined by the behavior of micro units and vice versa. CGE models link both spheres through a combination of microeconomic foundations and macroeconomic closure rules. In order to emulate the workings of the real economy as closely as possible, CGE models consist of two essential features. Firstly, they depict the behavior of economic agents and the functioning of markets through a set of simultaneous equations based on Walrasian general equilibrium theory. Secondly, they use a comprehensive and consistent dataset that feeds the mathematical model with actual data on production and consumption linkages.

The combination of the macro and micro sphere directly results from CGE models' theoretical foundation in Walrasian general equilibrium theory. The latter explains the functioning of the economy based on a bottom-up approach, where the individual welfare maximization behavior of rational producers and households determines the aggregated (macroeconomic) outcome (Sadoulet and De Janvry, 1995). Producers can perform different production activities that convert inputs (intermediate inputs and factors) into outputs. By maximizing their welfare, they choose to perform that production activity that will bring the highest profit, dependent on prices of all intermediate inputs, factors, and outputs. Households own production factors such as labor, land, and capital, which they sell in order to buy commodities for consumption. Subject to the value of their resource endowments and preferences, households choose an affordable consumption bundle of commodities that brings them the highest utility compared to all other bundles. A household's choice is dependent on the price of all commodities as well as on the prices of the factors owned by the household. Given these numerous decisions taken at the micro-level, an equilibrium of the whole economy requires a balance of supply and demand at all factor and product markets. In general equilibrium theory this is ensured through the price mechanism: There is a set of prices at which the decisions taken by all individual households and producers are compatible with each other. At these equilibrium prices, all commodity and factor markets clear and the equilibrium allocations of commodities and resources are both optimal and pareto-efficient.[4] Equilibrium thus means that total demand of any commodity equals the amount that is produced of this commodity, given available factors and inputs. At the same time, equilibrium prices postulate that all that is produced is demanded and that all factors available are employed in production (Wing, 2004).

The CGE models mirror general equilibrium theory through a system of linear and non-linear equations including production functions that define different production technologies and demand systems that define the demand behavior of households. The functional forms can differ between model formulations but are usually based on profit maximizing producers and utility maximizing households. Foreign trade in commodities is modeled endogenously, as producers can choose between supplying domestic markets and exporting, while consumers can demand imports or domestically produced commodities. To account for two-way trade of the same commodities, goods of foreign origin are imperfect substitutes to domestic goods (Diao and Thurlow, 2012). As national CGE models are quite large, the models do not include the behavior of every individual household or producer in an economy but work with aggregated representative agents, which might lead to losing important detail

4 The summary of general equilibrium theory is based on Hahn (1980).

(Böhringer et al., 2003). Apart from these behavioral equations, aggregated market clearing equations provide for equilibrium on all markets and endogenously solve for prices. Several closure rules ensure macroeconomic consistency and govern the behavior of the overall model. The closures can influence the impact of policies measures decisively and their specific functional form needs to be chosen individually as suitable for the country studied. The government is usually passive and its behavior is not modeled explicitly. Therefore, a government closure maintains a balanced budget between government income from taxes and expenditure for commodities and transfers. A savings-investment closure simply balances savings and investment. CGE models focus on the real economy and usually do not include financial or asset markets. Money is neutral and a price index is chosen as a numéraire, so that all prices are relative to this price index. The factor balance determines supply and mobility of production factors. Labor, land, and capital can be fully employed, i.e. limited, or unemployed. Capital is often sector specific, while land and labor are mobile and can move from less productive to more productive sectors (Sadoulet and De Janvry, 1995). Finally, the balance of payments closure regulates the current account. Models often assume a fixed current account deficit that is balanced by the real exchange rate. Changes in imports or exports then lead to exchange rate effects that affect relative prices of other commodities (Diao and Thurlow, 2012). The exact mathematical equations and suitable choice of closure rules for Malawi will be explained in more detail in section 2.3.2.1.

CGE models can in principle be applied to every country in the world if the mathematical model is calibrated with empirical data for the specific country. Model Parameters such as elasticities and variables such as consumption quantities must be computed with actual data to reflect the country-specific economic structure. The CGE model therefore includes a comprehensive database that mirrors the conditions of general equilibrium theory and forms the heart of the model. The social accounting matrix (SAM) is a consistent accounting framework where total expenditures match total income. A SAM encompasses all income and expenditure flows of a particular year within an economy between economic actors in the form of accounts and thus all production and consumption linkages (Diao and Thurlow, 2012). Most SAMs include different accounts for production activities, commodities, factors, households, the government, savings/investment and the rest of the world. Activities represent the producers or firms that produce commodities (goods and services). As one production activity is often able to produce different commodities, there are separate accounts for commodities and activities (Breisinger et al., 2009). The accounts are organized in a square matrix and each account includes a row for income and a column for expenditure. Table 2.1 shows an example of a typical aggregated (macro) SAM, where each cell is an expenditure from a column account to a row account. A SAM is consistent and complete as every flow of expenditure corresponds to a flow of income, column total must equal its respective row total (Dervis et al., 1982). The framework is completely flexible and each aggregated account as shown in Table 2.1 can be disaggregated if data is available. The higher the level of disaggregation the larger the data requirements. There is usually not one account for all commodities and activities respectively, but the main production sectors are represented in detail. Typical data sources include input-output tables that capture production linkages for activity and commodity accounts; nationally representative household surveys that provide information on consumption expenditure and factor endowments; government statistics for taxes and transfers; and national accounts providing information on total output, balance of payments, and trade (Sadoulet and De Janvry, 1995). The household account can be divided into different types of households, for example according to income quintile, or the factor account can be disaggregated into different labor types according to education. This disaggregation is important as it also determines the disaggregation of the model and its ability to measure distributional impacts. The latter is one of the great features of CGE models. Even though the CGE model works with representative household groups, detailed disaggregation of the household sector allows to analyze policy impacts on income and poverty especially for vulnerable household types.[5]

5 A similar discussion of SAMs was published by the author in Nielsen et al. (2015).

Table 2.1: Structure of an aggregated social accounting matrix (SAM)

	Activities	**Commod-ities**	**Factors**	**Enterprises**	**Households**	**Government**	**Investment**	**Rest of the World (RoW)**	**Total**
Activities		Marketed output			Home consumption				Activity income
Commodities	Intermediate inputs	Transaction costs			Marketed consumption	Government consumption	Investment, change in stocks	Exports	Total demand
Factors	Value-added							Foreign factor earnings	Factor earnings
Enterprises			Factor income to enterprises			Transfers to enterprises		Foreign enterprise receipts	Enterprise earnings
Households			Factor income to households	Indirect capital payments	Inter-household transfers	Transfers to households		Foreign remittances received	Household income
Government	Producer taxes	Sales taxes, import tariffs	Factor taxes	Corporate taxes	Personal taxes			Government transfers from RoW	Government income
Savings				Enterprise savings	Household savings	Government savings		Foreign savings	Savings
Rest of the World (RoW)		Imports		Repatriated earnings	Foreign remittances paid	Government transfers to RoW			Foreign exchange outflow
Total	Gross output	Total supply	Factor expenditure	Enterprise expenditure	Household expenditure	Government-expenditure	Investment	Foreign exchange inflow	

Source: Breisinger et al. (2009)

The SAM forms the base year from which policy simulations are conducted. Practically, a policy is simulated by exogenously shocking the model for example through changing the amount of factors available at certain markets. The impact of policies on an economy can be compared to distortions in the equilibrium of a market by affecting demand and supply, thereby changing the optimal allocation of resources. To take up the biofuel policy example from above, here an amount of land is exogenously given to biofuel production, which directly distorts the equilibrium on land markets. A new equilibrium of demand and supply in all markets is established through the price mechanism. The price and quantity adjustments by the "invisible hand" are also called general equilibrium effects and affect supply and demand of other sectors by changing prices and quantities at factor and product markets (Felderer and Homburg, 2005). Changes in resource endowments, an essential driver of nexus linkages, are thus simultaneously a driver of general equilibrium effects. At the new equilibrium established after the policy distortion, the circular flow of income is not interrupted, but the adjustments in different markets lead to changes in sectoral output and a new distribution of income among economic actors that can involve tradeoffs. There are two main groups of CGE models used for policy analysis: comparative static and dynamic recursive models. Static models are single period models and comprise the time span that it takes for the economy to reach equilibrium in all markets after a policy shock. They capture the medium term impacts of policy measures and usually cover the workings in an economy over several years or decades (Dervis et al., 1982). Static models compare the initial equilibrium with the final equilibrium, but do not consider the transition in between as well as the costs and benefits of this transition. Recursive dynamic models on the other hand take the transition from initial to final equilibrium into account by considering behavioral responses and may be able to measure potential tradeoffs in more detail (Piermartini and Teh, 2005). In recursive models, a sequence of equilibria is solved and parameters like capital accumulation and depreciation, productivity growth, factor and population growth are updated between the periods, which requires a lot of additional information and assumptions. Behavior of economic actors in each period depends on current prices assuming perfect information without considering forward-looking expectations (Diao and Thurlow, 2012). Recursive dynamic models and to some extent static models can thus account for important nexus drivers not only in terms of resource endowments but also concerning GDP and population growth. As the assumptions of CGE models about perfectly functioning markets, information and rational agents have been subject of critique, another class of general equilibrium models tries to overcome unrealistic assumptions. Dynamic stochastic general equilibrium (DSGE) models include uncertainty about parameters, expectations, and market imperfections, making them closer to real life. The computational requirements of modeling uncertainty are very large so that the economy is not represented in detail but at a high level of aggregation (Arora, 2013). Dynamic stochastic models are used to forecast macroeconomic variables such as economic growth and inflation and not distributional impacts of policy measures. There are other criticisms pertaining to CGE models, for example their sensitivity to choice of functional forms of equations for policy outcomes (McKitrick, 1998). Like all types of models, CGE models are no exact picture of reality, and actual impacts of policy measures might differ from simulation results. Notwithstanding, through their clear foundation in equilibrium theory the causal mechanisms can be easily traced and understood why a certain policy affects output and income in the way it does (Dervis et al., 1982). They are thus ideal to analyze and better understand the economic and social impacts of alternative policy measures on the FEW nexus in Malawi as explained in the next section.

2.2.4 CGE models' suitability for analyzing the FEW nexus in Malawi and the specific challenges

CGE models have long been used to analyze policies in a developing country context and their ability to trace both direct and indirect policy impacts stemming from economic linkages and market interactions is only one important feature (Dervis et al., 1982). Their other advantage lies in the flexibility of the structural form of equations and disaggregation to be applied to the special economic structure of the country studied. Although the functional forms of equations must be compatible with general equilibrium theory, certain equation systems especially in terms of macroeconomic closure rules can be chosen to mirror actual behavior and institutional arrangements (Shoven and Whalley, 1984). In the current account closure rule for example the exchange rate can be either fixed or floating depending on the monetary policy of the country. In a developing country context with a dominating agricultural sector and subsistence production, a focus on and thus detailed disaggregation of the rural economy in terms of agricultural activities and rural households is essential. As CGE models provide for this flexibility, they are an appropriate tool for quantitative policy analysis in Malawi.

CGE models can capture both the availability and access dimensions of the nexus. They track how policy measures affect allocation of resources, output and trade, measuring physical availability of commodities and resources. As the household sector can be disaggregated according to income quintiles, region, farm sizes, and even gender, the model can capture detailed distributional policy impacts in terms of income and identify the winners and losers. In addition, the access dimension of food, energy, and water security is measured through changes of real prices following policy interventions. Synergies and tradeoffs can thus be evaluated at the macro- and micro-level and for the economic and social dimensions of the FEW nexus. The special feature that CGE models encompass all market interactions is simultaneously a disadvantage as commodities that are not traded at markets are difficult to include. In terms of the nexus, this applies to subsistence production of food, biomass energy that is collected, as well as to water as a natural resource originating outside of markets. CGE models usually include subsistence production by assuming the same prices for purchased and home consumed commodities. As elaborated in section 2.2.1 this assumption might not be correct for semi-commercial farm households. CGE models are flexible enough that prices of home-produced and marketed commodities can differ, although this will increase required computing capacities. In certain functional forms of demand systems, consumption of home produced goods can be explicitly included. Moreover, even though households are treated as separate agents to producers in the CGE model, production and consumption decisions are not separated as all model equations are solved simultaneously.

2.2.4.1 Food

The CGE framework is well suited for analyzing food security in an agrarian developing country like Malawi. Many studies have explicitly and implicitly analyzed policy impacts on food security with CGE models such as Ecker et al. (2011) for Malawi and Diao et al. (2016) for Tanzania. While the CGE model cannot capture intra-household policy impacts in the way of farm household models, micro-simulation modules can be attached to CGE models to measure consumption and poverty changes in more detail. These modules are no behavioral models but accounting frameworks that contain the consumption expenditure of all households of an economy-wide household survey. The survey households are linked to their corresponding aggregated household group in the CGE model, and the consumption changes from the behavioral model are passed down to the survey households to recalculate consumption levels (Arndt et al., 2012). An increasing number of studies on Sub-Saharan African countries including Malawi has been linking CGE models with micro-simulation modules for poverty and nutrition (e.g. Arndt et al., 2008; Arndt et al., 2010b; Pauw and Thurlow, 2011; Pauw et al., 2012). In the poverty module, recalculated consumption levels of survey households are compared to the official poverty line of a country (Pauw et al., 2012). In nutrition modules, the initial food consumption of survey households is valued according to caloric availability of their diets. Food consumption changes from the CGE model are then used to recalculate changes in caloric availability of survey households and then compared to caloric requirements (Pauw and Thurlow, 2011). The nutrition micro-simulation module is the CGE models' attempt at measuring policy impacts on the utilization dimension of food security. Even though modules can measure changes in caloric availability, the results tell nothing about intra-household distribution of food or the preparation techniques employed. Utilization is therefore better analyzed with farm household models or microeconomic ex-post impact evaluation. The stability dimension of food security can be partly assessed by linking CGE models to stochastic weather models as will be explained in a later section.

2.2.4.2 Energy

Analyzing energy security in developing countries with CGE models remains challenging. One reason is that national accounts and trade data only contain formal energy supply such as electricity and traded fuels. In contrast, informal biomass energy in the form of firewood, charcoal, and crop residues dominates the energy sector in many countries in Sub-Saharan Africa, including Malawi. The energy supply base is thus given by the amount of available woods, and the supply of biomass energy is effectively an ecosystem service[6] provided by nature for free, if the wood does not come from plantations. Ecosystem services comprise environmental nexus linkages that cannot be directly captured with CGE models. Even if a price would be attached to the ecosystem service as is possible through different valuation techniques (see for example Pagiola et al., 2004), there is not any market interaction between nature and economic actors and thus no general equilibrium effect. The impact of policy measures on biomass supply from woods must thus be captured with biophysical models. As most biomass energy in rural areas is collected, energy production is part of subsistence production, but, unlike the consumption of home-produced food, usually not captured in economy-wide household surveys. This lack of data makes building

6 An ecosystem comprises the linkages between interacting species and their physical and biological environment and can differ in space and size (Alcamo et al., 2003). Ecosystem services are very broadly defined as the benefits provided by the ecosystem to people and include for example the provision of food and fuel but also soil formation, clean water and air through climatological processes (Alcamo et al., 2003).

comprehensive accounts for the energy sector in the SAM very difficult. If data on biomass energy consumption is known, an economy-wide shadow value can be attached to the amount of wood demanded. Most biomass energy is not traded at markets. The cost of collected biomass is therefore dependent on the time needed to collect firewood, and the respective household opportunity cost, given the collection time decreases the household's time for other activities (Fisher, 2004; Bandyopadhyay et al., 2011). The real value of energy for various households is thus very likely to differ immensely and cannot be equated with some economy-wide price. Even if markets exist for firewood and charcoal (mainly in urban areas), the prices at which biomass is purchased usually do not reflect prices at equilibrium demand and supply. People have to buy biomass in order to cook food and to survive and pay any price necessary. Surveys on charcoal in Tanzania and Malawi found that poorer households buy charcoal at higher prices and smaller quantities than richer households because they cannot afford to buy in bulk (Kambewa et al., 2007; World Bank, 2009).

CGE models alone are thus not well suited to analyze the biomass energy sector and micro-level energy security of households, which could be better achieved through a combination with farm household models. On the other hand, the availability dimension of energy security for formal energy supply such as biofuels and fossil fuels can well be measured. This is particularly useful when analyzing the impact of biofuel production policies on output, export, and prices of biofuels. Many developing countries promote biofuels to reduce their dependence on fossil fuel imports and to enhance domestic macro-level energy security (Arndt et al., 2010a). Studies have analyzed biofuel expansion policies with CGE models for Tanzania and Mozambique and implicitly considered nexus linkages. Arndt et al. (2010b) assess the economy-wide effects of increasing jatropha production in Mozambique on the availability of food. They find that an increase in energy output from biofuels may displace food crops due to insufficient infrastructure to access unused lands. Thurlow et al. (2015) analyze economic impacts of biofuel production in Tanzania combined with environmental impacts in the form of GHG emissions. The authors link the CGE model with a biophysical model to assess land use change and the consequences for emissions. Their findings indicate that higher energy security through biofuel production by smallholders negatively affects the environment as more lands have to be cleared than for large-scale production.

2.2.4.3 *Water*

As already mentioned, water is a natural resource that is difficult to comprehensively include into economic models, especially for developing countries. Water use of sectors is typically not considered in national accounts or input-output tables unless there is market for water, which is not the case for rural subsistence agriculture. Similar to biomass, water in developing countries is an ecosystem service that is not traded at markets and effectively a free resource. Formal water distribution systems usually only exist in urban areas for domestic and industrial use, while water in rural areas is collected. The shadow price of water in rural areas is thus very household specific or simply gratis in the case of rain water or if the household lives next to a water source. If limited water sources are shared, political economy costs can be substantial and affect the access to water. Empirical methods and farm household models can estimate individual shadow prices for rural household water and how policies affect micro-level water security (Aklilu, 2013). Rural water demand is not only affected by household decisions but also by environmental linkages that determine the water demand of crops. Evapotranspiration of both rainfed and irrigated crops is dependent on several climatological factors including sunshine, temperature, wind speed and humidity as well as crop specific physical and physiological features (Allen et al., 1998). Water demand from the agricultural sector can thus only be assessed with crop models that capture the impact of climate on evapotranspiration.

Water supply comes in the form of green water through rain and blue water from lakes, rivers and the ground. From a natural science perspective, the supply of water is endless considering the perpetual circle of evapotranspiration and rain. For a rainfed farmer, water might well be constrained especially during a drought. As water originates outside of markets, water supply cannot be captured with economic models. Water in rivers and lakes constantly flows and blue water availability cannot be measured at one point, but only over a space of time. Green water supply from rain can differ immensely in different years due to climate variability affecting the amount of available blue water and simultaneously water use and demand of crops. Global hydrology models that capture rainfall and runoff as well as water basin models that simulate the flows of blue water are needed to accurately assess water supply from various sources and feedback effects from policy measures (Robinson et al., 2015).

Although water supply requires an analysis with biophysical models, water demand has been included into CGE models. When formal water distribution systems and data on water use of different sectors exist, water can simply be added in the SAM as a separate account. It is treated as a commodity that is purchased by households for private consumption and by activities as a production factor (Calzadilla et al., 2011; Ponce et al., 2012). Through exogenously constraining water supply, the impact of climate change and water policies on production and households can be measured. Hassan and Thurlow (2011) for example built a Water-SAM for South Africa that includes industrial and domestic water demand using supply use data and household survey data. They also include irrigation water demand and estimate shadow prices for irrigation water for different crops based on experimental field data. This Water-SAM is included into a CGE model to measure impacts of water market policies (Hassan and Thurlow, 2011). On a global level Calzadilla et al. (2013) take the combination of CGE models with water a step further and explicitly model water supply and the environmental linkages between agriculture and water. They use an integrated modeling framework consisting of the global CGE model GTAP-W that includes water as a production factor, the above-mentioned agricultural sector IMPACT model and several hydrological models that capture changes in water availability due to climate change. Their modeling framework also encompasses feedback linkages of climate on water demand of crops and food production, capturing the interconnections between food and water security. The GTAP-W CGE model does not include data on water use outside of the agricultural sector and cannot capture micro-economic access to water. Nevertheless, the modeling framework emphasizes the importance of integrated modeling for a simultaneous assessment of nexus linkages and drivers, especially in terms of climate change.

2.2.4.4 Conclusion

This review has shown that CGE models are the most appropriate models to comprehensively assess economic nexus linkages and to identify policy tradeoffs and synergies at different levels of the economy. Even so, only the combination with biophysical models allows for a holistic analysis of the FEW nexus that includes environmental linkages. So far, no attempts have been made to explicitly study the nexus with a CGE based modeling framework. Since such a framework is essential to analyze policy impacts on the social and economic spheres of the FEW nexus, the studies in this work will close this important gap in research. As will be examined in more detail below, there are also cases where CGE models alone are not sufficient to analyze parts of the nexus so that completely new modeling approaches need to be developed, as in the case of the biomass energy sector. Overall, the correct modeling framework for policy analysis in the nexus depends on the intervention analyzed and the data at hand. As different policy measures affect different linkages, a partial view of the nexus in more detail rather than using a broad aggregated perspective might be necessary for analyzing specific policy impacts. The next section briefly discusses the environmental linkages that are important for the policy measures analyzed followed by the modeling frameworks developed for integrated nexus analysis.

2.3 Modeling the nexus: integrating the environment and non-market interactions

The previous section has given several examples of economic models combined with biophysical models to capture environmental linkages and ecological processes. There are a number of these processes that are particular important for the FEW nexus and the policies analyzed in the following chapters. Nature can be seen as an actor outside of the economy providing ecosystem services through water, land, and biomass that are essential for human survival and affect both the supply and demand of resources. Ecological processes are part of the earth's climate system including the atmosphere, hydrosphere, cryosphere, land cover and biosphere (Baede et al., 2001). The climate system influences nature's supply of water and growth of biomass including both trees and crops important for energy and food security. Growth of biomass is also influenced by the chemical and biological processes in soil. Where economic models omit environmental nexus linkages, crop models encompass these processes and capture how soil, water and the climate system affect the growth of plants (Doorenbos & Kassam, 1979; Jones et al., 2003).

In turn, human activities such as land clearing for agriculture influence the climate system and can inhibit the provision of ecosystem services. Land use change (LUC) influences the ability of ecosystems for soil carbon sequestration. If trees are cut, land clearing leads to an increase of GHG emissions in the atmosphere that can evoke global warming and climate change (Baede et al., 2001). On the other hand, man-made LUC can mean a decrease in emissions for example if agricultural land is planted with different crops that have higher carbon sequestration potential than the crops previously grown on the same land. LUC models can capture the impact of human activities on GHG emissions (e.g. Agarwal et al., 2002; Bernoux et al., 2011). Both crop and LUC models can be spatially explicit and analyses can be conducted for a specific country. As the whole climate system of the earth is involved in ecological processes, nature's role in the FEW nexus cannot be comprehensively included into policy analysis at an individual country level. The studies in this research therefore focus on specific environmental linkages that are affected by the policy measures analyzed. Before explaining the chosen modeling frameworks in detail, the next section briefly discusses which kind of models are suitable to capture global nexus linkages and could thus analyze the parts of the FEW nexus that national models cannot.

2.3.1 Integrated global models

Global systems analysis dates back to *The Limits to Growth* publication that was based on a world model that included socio-economic, technical and environmental linkages to capture the interplay between population growth, economic growth, pollution, and limited natural resources (Meadows et al., 1972). The simple world model evoked a need for more sophisticated global modeling that led to the establishment of the International Institute for Applied Systems Analysis (IIASA) that today maintains large global change models of both the environmental and economic sphere (Edwards, 1996). These models usually have a long-term temporal horizon of several decades and an emphasis on analyzing impacts of socio-economic drivers and policies on climate. The IIASA models can be integrated with each other and then form so called integrated assessment models (IAMs): for example the above-mentioned MESSAGE model, which only covers the energy supply side, is linked with a macroeconomic model to assess energy demand feedbacks or with an agricultural sector model to analyze impacts of energy policy on land, water, and forests (e.g. Messner and Schrattenholzer, 2000). In addition, the Agro-Ecological Zoning (AEZ) land use model from IIASA was combined with the above-mentioned water and energy models WEAP and the LEAP to explicitly analyze the FEW nexus in an integrated modeling framework. The Climate, Land, Energy and Water (CLEW) approach can quantify first round effects of policy measures on energy and water demand and supply, as well as changes in land use and output of different crops (Welsh et al., 2014). Integrated assessment models can thus be valuable tools for policy analysis in the FEW nexus on an aggregated level, but can only identify relatively broad policy impacts on different countries (Edwards, 1996).

Another common form of global change models are earth system models (ESM). ESMs are effectively natural science models that capture the earth's climate through modelling ecosystems and biogeochemical processes (Flato, 2011). They are usually linked to IAMs to include anthropogenic effects on climate and atmospheres. The Potsdam Institute for Climate Impact Research (PIK) for example uses a modeling combination of ESMs and IAMs, which has been applied to the FEW nexus to assess the impact of climate change mitigation on water demand for energy and food (Mouratiadou et al., 2016). Both model types are crucial for analyzing complex global systems, but are not suitable for analyzing the access dimension of food, energy, and water security.

2.3.2 An integrated modeling framework for nexus analysis in Malawi

In the subsequent chapters, four policy measures are analyzed that affect different nexus linkages. To capture the respective drivers and linkages in detail, each policy measure requires slightly different modeling frameworks. As CGE models encompass all economic linkages as well as resource endowments, population and economic growth, a CGE model applied to Malawi forms an inherent part of each modeling framework developed in this study.

2.3.2.1 Malawi CGE model

As discussed in the previous sections, the advantage of CGE models is their ability to contain the specific economic structure of developing countries within the bounds of general equilibrium theory. As a first step as part of this study, a new SAM was constructed for Malawi for the year 2010, using data from household surveys, supply-use tables, national accounts, government budgets, and balance of payments (NSO, 2012a; NSO, 2012b). Table 2.2 shows the aggregated macro SAM for Malawi for 2010, which is the most recent year for which an economy-wide household survey is available. As the data with which the SAM is compiled comes from different sources and usually does not completely add up, the SAM is balanced using cross entropy estimation (Robinson et al., 2001). The disaggregated SAM includes 59 activities and 50 commodities. Household and factor accounts were disaggregated using information from the 2010/2011 Integrated Household Survey (IHS) (NSO, 2012b). To measure distributional impacts on different households, the household sector is disaggregated into 30 representative groups according to rural and urban per capita expenditure quintiles and farm sizes (small, medium, large, non-farm). Land is disaggregated by farm size leading to four different land types (small, medium, large and estate), while labor is disaggregated by three education levels (not completed primary, primary and secondary education). Both disaggregations allow for a detailed analysis of policy impacts on different parts of the population. Capital is divided into agricultural and non-agricultural capital. The returns to the latter are first paid to a separate enterprise account and then transferred to households. All other factor incomes are directly paid to households by the factor accounts. The disaggregated SAM also contains tax accounts for direct (income) taxes and sales taxes that form the income of the government.

Table 2.2: 2010 Macro SAM for Malawi (millions of kwacha)

	Activities	Commodities	Factors	Enterprises	Households	Government	Investment	Rest of the World	Total
Activities		1,705,532							1,705,532
Commodities	733,923	147,937			918,696	162,955	169,863	183,696	2,317,071
Factors	971,609							68	971,677
Enterprises			240,808			28,515		35,792	305,114
Households			705,762	241,952		27,894		2,519	978,127
Government		87,338		56,767	47,750			48,368	240,223
Savings				6,102	9,437	18,109		136,215	169,863
Rest of the World		376,264	25,107	293	2,244	2,750			406,657
Total	1,705,532	2,317,071	971,677	305,114	978,127	240,223	169,863	406,657	

Source: Pauw et al. (2015)

The CGE model equations follow the IFPRI Standard Model described in Lofgren et al. (2002) and Diao and Thurlow (2012). In principle, the system of the model equations is formed by the behavior of welfare maximizing households and producers, constraints consisting of equilibrium conditions and the above-mentioned macroeconomic closures, foreign trade rules, and the government. In addition, several price equations link different endogenous prices (all domestic prices and factor wages) to other endogenous and exogenous prices (e.g. world market prices) and model variables. As the domestic price index is the model's numeraire, all endogenous and exogenous prices are relative to this price index. The functional forms reflect general equilibrium theory and are calibrated to Malawi's specific context. The most important equations are explained in the following, and all model equations for both the static and dynamic model can be found in detail in the appendix of Chapter 3.

In terms of household consumption in Malawi, the utility function must allow for non-monetary consumption of home produced goods as most rural households in Malawi practice subsistence farming. The demand functions of each representative household group are part of a linear expenditure system (LES) resulting from the maximization of a Stone-Geary-Utility function (Dervis et al., 1982, pp. 482–485).[7] The household consumption problem derives from the following mathematical form:

$$MAX\ \Pi_c\left(QH_{hc}-\gamma_{hc}\right)^{\beta_{hc}} \qquad (1)$$

$$\text{Subject to: } \Sigma_c\left(P_c * QH_c\right)=\left(1-sh_h-th_h\right)YH_h \qquad (2),$$

with $c\epsilon C$ and $h\epsilon H$ denoting different marketed or home produced commodities and households, respectively. γ is the so called subsistence minimum of a certain commodity. The subsistence minima exist for both marketed and home produced goods and can be seen as necessary quantities of different commodities that each household requires for living (Pollak and Wales, 1969). β represents the marginal budget share of a commodity and is calculated by multiplying the income elasticity for a certain commodity with its respective average budget share. β thus reflects the share of one monetary unit of income that is spend on a certain commodity. P_c are prices of different consumption quantities QH_c that depend on whether the commodity is purchased at market prices PQ_c (including marketing costs) or home produced. Prices of home produced goods PA_c correspond to their opportunity costs, which are their specific production activity costs. sh_h is the household's

7 Equations of the LES can be found in the appendix of chapter 5.

savings rate and th_h its tax rate, YH_h reflects total household income. Maximizing this utility function yields the following first order conditions or demand functions for m marketed commodities and hp home produced commodities.

$$PQ_c \cdot QH_{ch} = PQ_c \cdot \gamma_{ch}^{m} + \beta_{ch}^{m} \cdot \left(\left(1 - sh_h\right) \cdot \left(1 - th_h\right) \cdot YH_h - \sum_{c'} PQ_{c'} \cdot \gamma_{c'h}^{m} - \sum_{c'} PA_c \cdot \gamma_{ch}^{hp} \right) \quad (3)$$

$$PA_c \cdot QH_{ch} = PA_c \cdot \gamma_{ch}^{hp} + \beta_{ch}^{hp} \cdot \left(\left(1 - sh_h\right) \cdot \left(1 - th_h\right) \cdot YH_h - \sum_{c'} PQ_{c'} \cdot \gamma_{c'h}^{m} - \sum_{c'} PA_c \cdot \gamma_{ch}^{hp} \right) \quad (4)$$

The LES provides a complete system of demand functions and consumption of each commodity is dependent on prices of all other commodities, which reflects the assumptions of general equilibrium theory (Stone, 1954). This also means that all social and economic nexus linkages arising through household consumption are captured. In addition, the LES functional form allows (positive) nonunitary income elasticities of demand that can either be smaller than one for necessity goods or larger than unity for luxury goods (Diao and Thurlow, 2012). Both γ and β are dependent on each representative household's specific income elasticity to a certain commodity as well as average budget shares, which are econometrically estimated with data from the IHS. This means that each household group's demand decisions reflect its individual situation concerning income and home production, allowing for a detailed representation of distributional impacts of policy measures. An income increase might thus lead to an increase in food consumption of poorer households for whom food is a luxury good, whereas higher income quintiles may reduce their food consumption share as they see food as a necessity good. The features of the LES therefore not only conform to general equilibrium theory, but allow for a thorough measure of distributional impacts.

Similarly, the choice of production function reflects general equilibrium theory by enabling producers to substitute factors following changes in their relative prices. Equation (5) shows a constant elasticity of substitution (CES) function that differs for each sector's (activity's) production technology. $c \epsilon C$ again denote different commodities and $f \epsilon F$ different factors (land, labor, capital). QV_c stands for the output of a certain commodity using quantities of different factors QF_{fc} according to substitution elasticities ρ_c^p . α_c^p is an efficiency parameter reflecting total factor productivity and δ_{fc}^p a share parameter. Producers then maximize their profits employing factors until each factor's marginal revenue product equals the factor wage or rent. Intermediate inputs are used according to Leontief functions and fixed input-output coefficients that are calibrated through the data assembled in the SAM.

$$QV_c = \alpha_c^p \cdot \sum_f \left(\delta_{fc}^p \cdot QF_{fc}^{-\rho_c^p} \right)^{-1/\rho_c^p} \quad (5)$$

As already mentioned imported commodities are treated as imperfect substitutes to the same domestically produced commodities to explain the phenomenon of two-way trade and different consumer preferences. This imperfect substitutability is defined by a CES function known as the Armington function, who first introduced this solution (Armington, 1969). Similarly, producers take the decision to export based on a constant elasticity of transformation function. Both equations can be found in the annex of Chapter 3. Substitution between domestic and foreign goods is governed by the relative prices of those goods and substitution elasticities from Dimaranan (2006). Since Malawi is a small country, changes in Malawi's output or consumption do not have any influence on world markets, therefore world market prices for imports and exports are fixed.

Several equilibrium conditions ensure that all markets clear. Firstly, total factor supply equals total factor demand. As Malawi faces high land and labor constraints during peak production times in the rainy season, all factors are assumed to be fully employed. As will be shown in Chapter 4, this assumption might not always be correct. Land and labor are mobile across sectors, while capital is sector-specific. In addition, household income equals the returns from factors (minus taxes) subject to their factor endowments and potential transfers such as remittances. Households that are richer in terms of land or skilled labor thus usually receive a larger share of factor income. Overall equilibrium is governed by the following equation that defines balance between total supply and total demand.

$$QQ_c = \sum_{c'} ca_{cc'} \cdot QN_{c'} + \sum_h QH_{ch} + QG_c + QI_c + QT_c \quad (6)$$

The composite supply quantity of all domestically produced goods and imports QQ_c has to be equal to all domestically consumed goods including aggregate intermediate inputs QN_c (multiplied with the input coefficients *ca*), aggregate household consumption QH_{ch}, government consumption QG_c, investment demand QI_c and transaction and trade commodities QT_c.

This overall equilibrium condition is directly related to the three macroeconomic closure rules. Government behavior is passive and government revenue is defined by the sum of all taxes and foreign transfers such as foreign aid. Government expenditure consists of consumption of commodities such as services for education and health as well as the recurrent fiscal balance. Under the government closure, government tax rates and recurrent spending are fixed in the model for Malawi and the recurrent balance (usually at deficit) adjusts to balance total revenues and expenditures. The level of investment demand is balanced through the savings-investment closure, where the sum of private, government, and foreign savings must equal total investment demand. Private savings rates are fixed and investment is savings-driven – when savings increase due to higher income, so does investment. Such an assumption makes sense as private savings and investment are extremely small in Malawi and all significant investment must come from abroad. Finally, the most important closure is the current account balance, which is given below. It postulates balance between Malawi's foreign exchange income and expenditure and represents Malawi's flexible exchange rate regime. Foreign spending consists of consumption of imports at fixed world market prices as well as factor transfers to the rest of the world (factor income YF_f multiplied with foreign remittance rate rf_f minus direct taxes tf_f). The latter are denoted in domestic prices and need to be multiplied by the exchange rate X. Foreign exchange income is derived from selling exports QE_c at fixed world market prices pwe_c as well as from foreign net transfers wh_h such as remittances. cab represents a fixed current account deficit, while Malawi's flexible exchange rate balances the current account closure.

$$\begin{aligned}&\sum\nolimits_c pwm_c \cdot QM_c + \sum\nolimits_f \left(1 - tf_f\right) \cdot rf_f \cdot YF_f \cdot X^{-1} = \\ &\sum\nolimits_c pwe_c \cdot QE_c + \sum\nolimits_h wh_h + cab\end{aligned} \qquad (7)$$

The just described equations lie at the heart of the CGE model of Malawi that forms the basis of all three studies in this book. The following sections explain the newly developed modeling frameworks for analyzing the FEW nexus beyond the linkages included in the CGE model.

2.3.2.2 Modeling framework for analyzing biofuel production

The expansion of biofuel production triggers several nexus linkages that need to be captured in a thorough assessment of policy impacts. Figure 2.2 illustrates the modeling framework developed for assessing the economic, social, and environmental spheres affected by biofuel production. Biofuel crops are often blamed for decreasing food security by displacing food crops and competing for resources, but are also found to increase welfare of biofuel crop producing households. The CGE model linked with a micro-simulation poverty module encompasses this economic and social sphere of biofuel expansion. Effects on macro-level energy and food security are directly measured by output of biofuels and food crops. Both the CGE model and poverty module assess impacts of biofuel production on the access dimension of security. As both water and biomass energy are not traded at markets except for in some urban areas, household energy and water security cannot be analyzed with these models. However, overall effects on poverty and income shed light on welfare effects of biofuel expansion. Food security impacts are captured by consumption changes of households.

Figure 2.2: Modeling framework for analyzing biofuel production

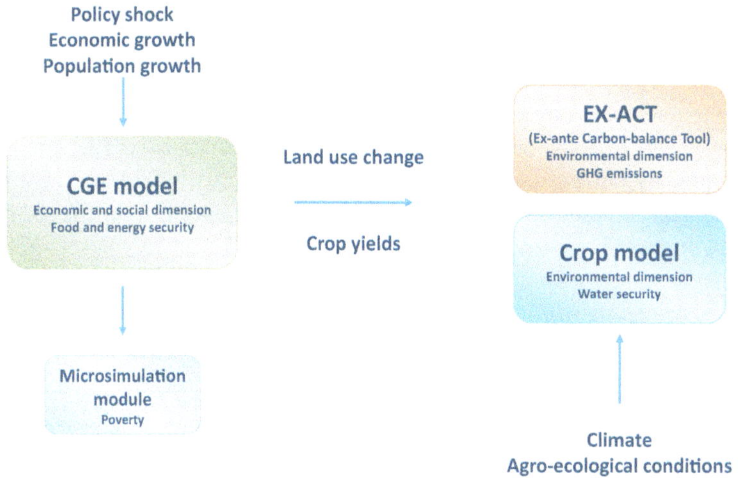

Source: Author's creation

The environmental sphere of biofuel production is analyzed through two models that are linked with the CGE model. Land clearing and land use change for biofuels can increase GHG emissions, which can further accelerate global warming and climate change. The ex-ante carbon balance tool (EX-ACT) measures changes in GHG emissions following land use changes (Bernoux et al., 2011). The latter are an output of the CGE model that is fed into EX-ACT to determine how biofuel production affects GHG emissions in Malawi. An important assumption of the modeling framework is that land clearing only happens on grasslands. This has implications for both GHG emissions and energy security, as no additional trees are cut that fulfil important ecosystem services in terms of soil carbon sequestration and biomass energy. Therefore, there is no negative impact on access to biomass energy for households. At the same time, climate and the overall agro-ecological conditions in Malawi affect how biofuel crops grow and how much water Malawian crops require. Biofuel crops, especially sugarcane, require a lot of water and are often irrigated. The ecological processes around crop growth and the impact of climate as well as irrigation on crop water use in Malawi are measured by a newly developed crop model based on a yield response to water approach (Doorenbos and Kassam, 1979). The CGE results in terms of crop production are passed down to this crop model to assess water security impacts of rainfed and irrigated biofuel production.

This integrated modeling framework allows to simultaneously assess economic, social, and environmental impacts of biofuel expansion on the FEW nexus in Malawi. The model is run over a ten year period, in which nexus drivers in the form of population and economic growth influence the effectiveness of biofuel expansion regarding food, energy, and water security. The framework is thus well suited to analyze the overarching research questions posed in Chapter 1 as well as the specific research questions pertaining food security, water use, and GHG emissions. By encompassing all nexus linkages affected by biofuel expansion, the modeling framework is able to assess the synergies and tradeoffs for food, energy, and water that are generated by this policy.

2.3.2.3 Modeling framework for analyzing irrigation expansion

While the previous modeling framework is ideal to assess impacts of increased biofuel production on the FEW nexus under rainfed and irrigated conditions, it is not able to capture the environmental linkages around irrigation and crop growth in detail. The assessment of irrigation expansion in Malawi therefore requires a slightly different modeling approach compared to the above framework as other nexus linkages are affected and is shown in Figure 2.3. Economic and social linkages are again analyzed with the Malawi CGE model, which is combined with a poverty module that captures distributional impacts on survey households in more detail. The impact of irrigation on the economy and food security depends on the one hand on its potential in increasing crop yields and therefore output. On the other hand, the distributional impacts of irrigation depend on who benefits from increased yields, especially since production costs are likely to increase compared to rainfed production. The distributional effects are largely dependent on the actual costs of irrigation for the individual farmer in terms of labor and the effect on output and prices, all of which can be captured with the CGE model framework. To analyze the impact of irrigation on crop yields, the CGE model is again linked to a crop model, albeit a more sophisticated crop model as used in the previous modeling framework. The role of the crop model is not to determine water use of crops per se, but rather to estimate how irrigation affects crop growth considering changes in climate, agro-ecological conditions as well as management techniques. The Decision Support Software for Agrotechnology Transfer (DSSAT) cropping system model simulates the growth of plants and calculates yield effects under various crop management techniques (e.g. fertilizer or irrigation) using daily weather information to assess changes in soil water and nutrients (Jones et al., 2003). Apart from overall yield effects, the crop model sheds light on how ecological processes such as the interaction between water and nutrients determine crop growth and subsequent economic impacts. The model thus captures environmental nexus linkages in terms of water, soil, and climate affecting economic growth and food security.

Figure 2.3: Modeling framework for analyzing irrigation expansion

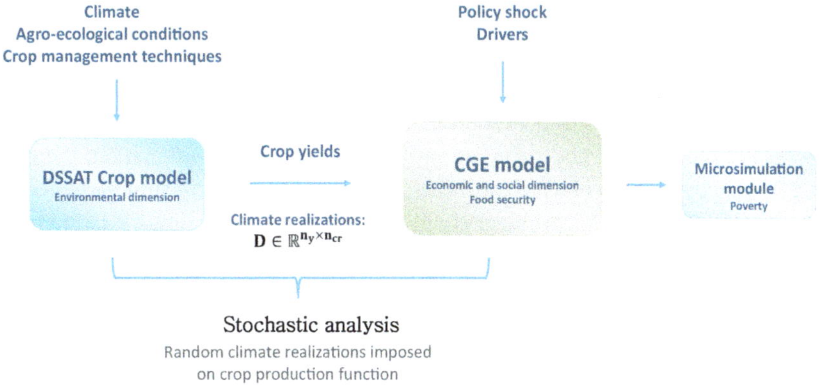

Source: Author's creation

Irrigation not only increases yields, but also affects the resilience of crops to climate variability and therefore the vulnerability of farmers and the whole economy to climate change, one of the most important nexus drivers. To capture the effect of irrigation on reducing risks and vulnerability to climate change, a stochastic component is added to the crop and CGE model. Historic climate realizations are imposed on the crop model, which translates variation in climate into the corresponding yield realizations of crops in Malawi under different management techniques. These realizations form a historic yield distribution, from which random historic climate events are drawn and imposed on the CGE model to measure the risk-reducing potential of irrigation for the economy as a whole and resource-poor farmers. The stochastic simulations also allow to assess the stability dimension of food security under climate variability.

Macro-level energy and water security are implicitly incorporated in the modeling approach as the irrigation potential simulated considers competing water uses for humans, the environment and hydroenergy production. The modeling framework thus not only encompasses the simultaneous assessment of the FEW nexus components and distributional impacts, but also includes the role of climate change in the effectiveness of irrigation. It is thus well suited to capture potential synergies and tradeoffs of irrigation. Although household level energy security cannot be assessed with this framework, no land clearing is undertaken in the scenarios that would affect the availability of biomass energy. Increases in crop yields and welfare will also increase demand for energy, making it imperative to find policy measures that can increase biomass energy supply even in the light of economic and population growth.

2.3.2.4 Modeling framework for analyzing the biomass energy sector

There are several reasons why the biomass energy sector in Malawi cannot be adequately modeled with a CGE model. First of all, basically all wood in rural areas is collected and not traded at markets. Where there is no market, there is no market price, both of which are central to the general equilibrium mechanism of the CGE model. This problem could in principle be overcome by estimating household shadow prices for collected fuelwood as it is done by econometric studies (e.g. Chen et al., 2006; Murphy et al., 2015). Shadow prices are usually calculated based on the opportunity costs of time for collection, which can be the off-farm labor wage or specific shadow wages as estimated in Murphy et al. (2015). As firewood is predominantly collected by women and children, who do not face many off-farm labor possibilities, their time opportunity costs for collection is unlikely to mirror the wage rate. In addition, opportunity costs are likely to differ immensely across households, making the shadow price of fuelwood very household specific. Another alternative to estimate prices could be to use urban charcoal and firewood market prices for the whole economy. Even if one of these methods would yield prices that are adequate for economy-wide modeling, rural biomass supply as a service provided by nature is independent from market considerations. As long as fuelwood is not grown by humans on plantations or through reforestation, there is no economic supply response by nature to increases or decreases in demand. In addition, there is no recent nor reliable data on biomass energy demand in Malawi to date.

Given these problems in including the biomass energy sector in a CGE model per se, innovative modeling approaches are developed that are linked to the CGE model as shown in Figure 2.4. The CGE model captures all economic and social linkages of the nexus and the drivers economic and population growth. A newly developed food-energy model uses the direct linkage between food and energy to estimate cooking energy demand from food demand. In order to assess the impact of the two drivers and the rest of the economy on biomass energy demand, the model is linked to the CGE model so that food demand changes due to economic and population growth are translated into energy demand. The food-energy model than allows for analyzing the impact of demand side policies such as cooking efficiency increases through adoption of improved cookstoves. The supply side is on the one hand assessed through a supply module that captures the environmental dimension by estimating sustainable supply based on land cover and sustainable biomass yields. On the other hand, analyzing the impact of supply side policies in the form of agroforestry requires an additional tailor-made model. As households have to practice agroforestry on their fields, the willingness and decision of households to undertake additional work on the farm is essential for the success of increasing biomass supply. Therefore, a farm household model is developed that considers the individual household's preferences and endowments in terms of land and labor, thereby encompassing the social dimension of the FEW nexus as well.

Figure 2.4: Modeling framework for analyzing the biomass energy sector

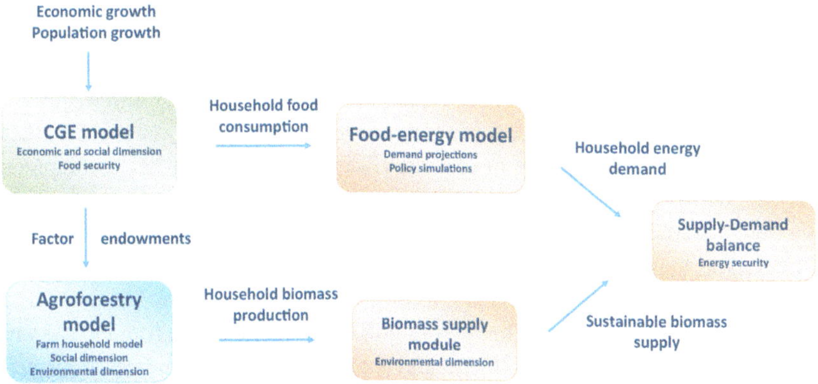

Source: Author's creation

Such a detailed assessment of the biomass energy sector and policy impacts leads to tradeoffs in terms of modeling. The framework used does not attempt to model the complete nexus by including water and climate in the analysis. Nevertheless, the policy measures analyzed evoke no tradeoffs for the environment. On the contrary, agroforestry and reduced fuel needs through improved cookstoves will have positive impacts on micro-climate and watersheds, as trees increase the soil's ability to hold water and reduce run-off (Malmer et al., 2010). The emphasis of this modeling framework is put on capturing the relevant economic and environmental linkages between food and energy security that are affected by the policy measures analyzed. At the same time, the framework assesses the potential for a sustainable biomass energy sector under the pressure of the drivers economic and population growth, as sustainable development is essential to improve the livelihoods of the poorest in Malawi.

2.4 Conclusion

This chapter has explained in detail the role of economic and social linkages in the FEW nexus. The review of different economic modeling approaches has shown that only CGE models are able to capture all economic linkages and indirect effects of policy measures on the whole economy. If nexus linkages occur outside of markets or the success of policies is dependent on intra-household decision making processes, different approaches such as farm household models can better simulate potential policy impacts. To simultaneously assess environmental nexus linkages, economic models need to be combined with biophysical models, especially crop models that capture the interactions between climate, water, soil, and crop growth. An important conclusion is that no modeling framework is perfect in a sense that all economic and ecological linkages between food, energy, and water can be modeled. The interconnections are so complex and manifold, especially in terms of ecological processes around climate and atmosphere, that any attempt at modeling the nexus will always remain somewhat incomplete. As different policy measures trigger different nexus linkages, a more focused modeling framework does not imply any disadvantages but can help to better understand the effectiveness of policies.

So far no attempt has been made to explicitly analyze the FEW nexus with a CGE based suite of models. Three types of integrated modeling frameworks are developed to close this research gap and assess the impact of increased biofuel production, irrigation expansion, improved cookstoves and agroforestry on the FEW nexus in Malawi. The modeling frameworks concentrate on simultaneously capturing the relevant economic and environmental linkages that are affected by the respective policy measures analyzed. By fulfilling the methodological objective of this study, not only the research questions posed in the previous chapter are answered, but also the empirical objective of identifying those interventions that maximize synergies and minimize tradeoffs in the food-energy-water nexus is achieved.

References

Alcamo, J., et al., 2003. Ecosystems and human well-being: a framework for assessment/Millennium Ecosystem Assessment. Island Press, Washington DC, USA.

Agarwal, C., Green, G.M., Grove, J.M., Evans, T.P., Schweik, C.M., 2002. A Review and Assessment of Land-Use Change Models: Dynamics of Space, Time, and Human Choice Gen. Tech. Rep. NE-297. Newtown Square, PA: U.S. Department of Agriculture, Forest Service, Northeastern Research Station.

Allen, R.G., L.S. Pereira, D. Raes, Smith, M., 1998. Crop Evapotranspiration – Guidelines for Computing Crop Water Requirements. *Irrigation and Drainage paper 56*, United Nations Food and Agriculture Organization, Rome, Italy.

Alwang, J., Siegel, P.B., 1999. Labor Shortages on Small Landholdings in Malawi: Implications for Policy Reforms. *World Development*, 27(8), 1461–1475.

Aklilu, A., 2013. Water, Smallholders and Food Security – An econometric assessment of the effect of time spent on collecting water on households' economy and food security in rural Ethiopia. Master's thesis, Environmental Economics and Management Master's Programme, Faculty of Natural Resources and Agricultural Sciences, Department of Economics, Swedish University of Agricultural Sciences, Uppsala, Sweden.

Armington, P., 1969. A Theory of Demand for Products Distinguished by Place of Production. Staff Papers (International Monetary Fund), 16(1), 159–178.

Arndt, C., Benfica, R., Tarp, F. Maximiano, N., Nucifora, A.M.D. & Thurlow, J., 2008. Higher Fuel and Food Prices, Economic, Impacts and Responses for Mozambique. IFPRI Discussion Paper 00836.

Arndt, C., Msangi, S., & Thurlow, J. 2010a. Are Biofuels Good for African Development? An Analytical Framework with Evidence from Mozambique and Tanzania. UNU-WIDER Working Paper No. 2010/110.

Arndt, C., Benfica, R., Tarv, F., Thurlow, J., Uaiene, R., 2010b. Biofuels, poverty, and growth: a computable general equilibrium analysis of Mozambique. *Environmental and Development Economics*, 15, 81–105.

Arndt, C., Pauw, K., Thurlow, J., 2012. Biofuels and economic development: A computable general equilibrium analysis for Tanzania. *Energy Economics*, 34(6), 1922–1930.

Arora, V., 2013. An Evaluation of Macroeconomic Models for Use at EIA. U.S. Energy Information Administration Working Paper Series.

Arulpragasam, J., Conway, P., 2003. Partial Equilibrium Multi-market Analysis. In Bourguignon, F., L. Pereira da Silva, L., (eds.): The Impact of Economic

Policies on Poverty and Income Distribution. Oxford, UK: Oxford University Press.

Baede, A.P.M., Ahlonsou, E., Ding, Y., Schimel, D., 2001. The Climate System: An Overview. In: Climate Change 2001: The Scientific Basis. Contribution of Working Group I to the Third Assessment Report of the Intergovernmental Panel on Climate Change [Houghton, J.T., Y. Ding, D.J. Griggs, M. Noguer, P.J. van der Linden, X. Dai, K. Maskell, and C.A. Johnson (eds.)]. Cambridge University Press, Cambridge, United Kingdom and New York, NY, USA.

Bandyopadhyay, S., Shyamsundar, P. & Baccini, A., 2011. Forests, biomass use and poverty in Malawi. *Ecological Economics*, 70(12), 2461–2471.

Bazilian, M., Rogner, H., Howells, M., Hermann, S., Arent, D., Gielen, D., Steduto, P., Mueller, A., Komor, P., Tol, R.S.J. & Yumkella, K. K., 2011. Considering the energy, water and food nexus: Towards an integrated modelling approach. Energy Policy, 39(12), 7896–7906.

Beneria, L., 1979. Reproduction, production and the sexual division of labor. *Cambridge Journal of Economics*, 3, 203–225.

Bernoux, M., Tinlot, M., Bockel, L., Branca, G., Gentien, A., 2011. *EX-ante carbon-balance tool (EX-ACT): technical guidelines for Version 3, Easypol module 101.* Food and Agriculture Organization of the United Nations, Rome, Italy.

Bhattacharyya, S., Timilsina, G., 2010. A review of energy system models. International Journal of Energy Sector Management, Vol. 4 (4), pp. 494–518.

Blanco-Fonseca, M., 2010. Literature Review of Methodologies to Generate Baselines for Agriculture and Land Use. WP4 Baseline, Deliverable: D4.1, Common Agricultural Policy Regional Impact – The Rural Development Dimension Collaborative project – Small to medium-scale focused research project under the Seventh Framework Programme, Project No.: 226195, European Commission – Joint Research Centre (JRC).

Böhringer, C., Rutherford, T.F., Wiegard, W., 2003. Computable General Equilibrium Analysis: Opening a Black Box, Discussion Paper No. 03–56, Center for European Economic Research, Mannheim, Germany.

Breisinger, C., Thomas, M. & Thurlow, J., 2009. Social accounting matrices and multiplier analysis: An introduction with exercises. Food Security in Practice technical guide 5. Washington, D.C.: International Policy Research Institute.

Calzadilla, A., Rehdanz, K., Tol, R.S.J., 2011. Water scarcity and the impact of improved irrigation management: a computable general equilibrium analysis. *Agricultural Economics*, 42(3), pp. 305–323.

Calzadilla, A., Zhu, T., Rehdanz, K., Tol, R.S.J., Ringler, C., 2013. Economywide impacts of climate change on agriculture in Sub-Saharan Africa. *Ecological Economics*, 93, 150–165.

Chen, L., Heerink, N., Van Den Berg, M., 2006. Energy consumption in rural China: A household model for three villages in Jiangxi Province. *Ecological Economics*, 58, 407–420.

Dervis, K., J. De Melo, and S. Robinson, 1982. General Equilibrium Models for Development Policy. New York: Cambridge University Press.

Diao, X., Thurlow, J., 2012. A Recursive Dynamic Computable General Equilibrium Model, in X. Diao, J. Thurlow, S. Benin and S. Fan, eds., *Strategies and Priorities for African Agriculture: Economywide Perspectives from Country Studies*. International Food Policy Research Institute (IFPRI), Washington, DC.

Diao, X., Kennedy, A., 2016. Economywide Impact of Maize Export Bans on Agricultural Growth and Household Welfare in Tanzania: A Dynamic Computable General Equilibrium Model Analysis. *Development Policy Review*, 34(1), 101–134.

Doorenbos, J., Kassam, A.H., 1979. Yield Response to water. Irrigation and Drainage paper 33, United Nations Food and Agriculture Organization, Rome, Italy.

Dimaranan, Betina V., Editor (2006). Global Trade, Assistance, and Production: The GTAP 6 Data Base, Center for Global Trade Analysis, Purdue University.

Ecker, O., Breisinger, C. & Pauw, K., 2011: Growth is good but is not enough to improve nutrition. 2020 Conference: Leveraging Agriculture for Improving Nutrition and Health, February 10–12, 2011, New Delhi, India. 2020 Conference Paper 7.

Edwards, P., 1996. Global Comprehensive Models in Politics and Policymaking, Editorial Essay. *Climate Change*, 32, 149–161.

Felderer, B. & Homburg, S., 2005. Makroökonomik und neue Makroökonomik, Springer: Berlin, Heidelberg.

Fisher, M., 2004. Household welfare and forest dependence in Southern Malawi. *Environment and Development Economics*, 9, 135–154.

Flato, G., 2011. Earth system models: an overview. *Wiley Interdisciplinary Reviews: Climate Change*, 2(6), 783–800.

Hahn, F., 1980. General Equilibrium Theory. *The Public Interest*, 58 (special issue), 123–138.

Hassan, R., & Thurlow, J. (2011). Macro-micro feedback links of water management in South Africa: CGE analyses of selected policy regimes. *Agricultural Economics*, 42(2), 235–247.

IEA-ETSAP, 2014. MARKAL. http://www.iea-etsap.org/web/Markal.asp, access date: 10-01-2014.

IIASA, 2014. MESSAGE. http://www.iiasa.ac.at/web/home/research/research-Programs/Energy/MESSAGE.en.html, access date: 10-01-2014.

Jones J.W., G. Hoogenboom, C.H. Porter, K.J. Boote, W.D. Batchelor, L.A. Hunt, P.W. Wilkens, U. Singh, A.J. Gijsman, and J.T. Ritchie. 2003. The DSSAT cropping system model. European Journal of Agronomy, 18(3–4): 235–265.

Kambewa, P., Mataya, B., Sichinga, W., Johnson, T., 2007. Charcoal: the reality – a study of charcoal consumption, trade and production in Malawi. Small and Medium Forestry Enterprise Series, 21. London: International Institute for Environment and Development.

Lofgren H., R.L. Harris and S. Robinson. 2002. A Standard Computable General Equilibrium (CGE) Model in GAMS. Washington D.C.: IFPRI.

Malmer, A., Scott, D., Vignola, R., Xu, J., 2010. Forest cover and global water governance. In: Forests and Society – Responding to Global Drivers of Change (eds Mery, G et al.), pp. 75–93. IUFRO World series, No. 25, International Union of Forest Researchers Organization (IUFRO), Vienna.

McKitrick, R.R., 1998. The econometric critique of computable general equilibrium modeling: the role of functional forms. *Economic Modelling*, 15, 543–573.

Meadows, D. H., Meadows, D. L., Randers, J., Behrens III, W. W., 1972. The Limits to Growth: a report for the Club of Rome's project on the predicament of mankind. Potomac Associates, Washington DC, USA.

Mekonnen, D., Bryan, E., Alemu, T., Ringler, C., 2015. Food versus Fuel: Examining Tradeoffs in the Allocation of Biomass Energy Sources to Domestic and Productive Uses in Ethiopia. Selected Paper prepared for presentation at the 2015 Agricultural & Applied Economics Association and Western Agricultural Economics Association Annual Meeting, San Francisco, CA, July 26-28.

Messner, S., Schrattenholzer, L., 2000. MESSAGE – MACRO: linking an energy supply model with a macroeconomic module and solving it iteratively. *Energy*, 25, 267–282.

Mouratiadou, I., Biewald, A., Pehl, M., Bonsch, M., Baumstark, L., Klein, D., Popp, A., Luderer, G., Kriegler, E., 2016. The impact of climate change mitigation on water demand for energy and food: An integrated analysis based on the Shared Socioeconomic Pathways, *Environmental Science and Policy*, 64, 48–58.

Murphy, D., Berazneva, J., Lee, D.R., 2015. Fuelwood Source Substitution and Shadow Prices in Western Kenya. Selected Paper prepared for presentation at the 2015 Agricultural & Applied Economics Association and Western Agricultural Economics Association Annual Meeting, San Francisco, CA, July 26–28.

Njewa, E., 2012. The Road Towards Low Carbon Development Strategy in Malawi. Presentation held at UNFCCC Workshop, May 18, Bonn Climate Change Conference, Bonn, Germany.

NSO (National Statistical Office), 2012a. Statistical Yearbook 2012. Zomba, Malawi.

NSO (National Statistical Office), 2012b. Integrated Household Survey 2010/11. Lilongwe, Malawi.

Pagiola, S., v. Ritter, K., Bishop, J., 2004. Assessing the Economic Value of Ecosystem Conservation. World Bank Environment Department Paper No. 101, The World Bank, Washington DC, USA.

Pattanayak, S., Sills, E., Kramer, R., 2004. Seeing the forest for the fuel. *Environment and Development Economics*, 9,155–179.

Pauw, K., & Thurlow, J., 2011. Agricultural growth, poverty, and nutrition in Tanzania. *Food Policy*, 36(6), 795–804.

Pauw, K., Thurlow, J., Uaiene, R. & Mazunda, J., 2012. Agricultural Growth and Poverty in Mozambique: Technical Analysis in Support of the Comprehensive Africa Agriculture Development Program (CAADP). Mozambique Strategy Support Program, Working Paper 2.

Pauw, K., Schuenemann, F., Thurlow, J., 2015. A 2010 Social Accounting Matrix for Malawi. Unpublished mimeograph, International Food Policy Research Institute, Washington DC, USA.

Piermartini, R.; Teh, R., 2005. Demystifying Modelling Methods for Trade Policy. WTO Discussion Paper 10 (September), World Trade Organization, Geneva.

Pollak, R. A., Wales, T. J., 1969. Estimation of the Linear Expenditure System. *Econometrica*, 37(4), pp. 611–628.

Ponce, R., Bosello, F., & Giupponi, C., 2012. Integrating Water Resources into Computable General Equilibrium Models – A Survey. Fondazione Eni Enrico Mattei Working Paper Series 57.2012.

Robinson, S., Cattaneo, A., El-Said, M. 2001. "Updating and Estimating a Social Accounting Matrix Using Cross Entropy Methods," *Economic Systems Research*, 13(1), 47–64.

Robinson, S., Mason d'Croz, D., Islam, S., Sulser, T. B., Robertson, R. D., Zhu, T., Gueneau, A., Pitois, G., Rosegrant, M. W., 2015. The International Model for Policy Analysis of Agricultural Commodities and Trade (IMPACT): Model description for version 3. IFPRI Discussion Paper 1483. Washington, D.C.: International Food Policy Research Institute (IFPRI).

Sadoulet, E., De Janvry, A., 1995. Quantitative development policy analysis. Johns Hopkins University Press, Baltimore and London.

SEI, 2013. LEAP brochure, http://www.sei-us.org/Publications_PDF/SEI-LEAP-brochure-Jan2012.pdf, access date: 10-01-2014.

Shoven, J., Whalley, J., 1984. Applied General-Equilibrium Models of Taxation and International Trade: An Introduction and Survey. *Journal of Economic Literature*, 22(3), 1007–1051.

Singh, I., Squire, L., Strauss, J. [editors], 1986. Agricultural household models: extensions, applications, and policy. Baltimore, MD: The Johns Hopkins University Press.

Stone, R., 1954. Linear Expenditure Systems and Demand Analysis: An Application to the Pattern of British Demand. *The Economic Journal*, 64(255), 511–527.

Taylor, J., Adelman, I., 2003. Agricultural Household Models: Genesis, Evolution, and Extensions. *Review of Economics of the Household*, 1(1), 33–58.

UNEP, 1999. Climate Change Mitigation Study in Southern Africa, Tanzania Country Study. UNEP Collaborating Centre on Energy and Environment, Risø National Laboratory, Denmark.

Van Wijk, M.T., Rufino, M.C., Enahoro, D., Parsons, D., Silvestri, S., Valdivia, R.O., Herrero, M., 2014. Farm household models to analyse food security in a changing climate: A review. *Global Food Security*, 3, 77–84.

WEAP21, 2014. Water Evaluation and Planning. http://www.weap21.org/index.asp, access date: 10-01-2014.

Welsh, M., Hermann, S., Howells, M., Rogner, H. H., Young, C., Ramma, I., Bazilian, M., Fischer, G., Alfstad, T., Gielen, D., Le Blanc, D., Röhrl, A., Steduto, P. and A. Müller, 2014. Adding value with CLEWS – Modelling the energy system and its interdependencies for Mauritius. *Applied Energy*, 113, 1434–1445.

Wing, I.S., 2004. Computable General Equilibrium Models and Their Use in Economy-Wide Policy Analysis. MIT Joint Program on the Science and Policy of Global Change, Technical Note No. 6.

World Bank, 2009. Environmental Crisis or Sustainable Development Opportunity? Transforming the charcoal sector in Tanzania. A Policy Note. Washington DC, USA.

3 Leveling the field for biofuels: comparing the economic and environmental impacts of biofuel and other export crops in Malawi[1]

Abstract: Biofuels often raise the specter of food insecurity, water resource depletion, and greenhouse gas emissions from land clearing. These concerns underpin the "sustainability criteria" governing access to European biofuel markets. However, it is unclear if producing biofuels in low-income countries does exacerbate poverty and food insecurity, and moreover, whether the sustainability criteria should apply to all agricultural exports entering European markets. We develop an integrated modeling framework to simultaneously assess the economic and environmental impacts of producing biofuels in Malawi. We incorporate the effects of land use change on crop water use, and the opportunity costs of using scarce resources for biofuels instead of other crops. We find that biofuel production reduces poverty and food insecurity by raising household incomes. Irrigated outgrower schemes, rather than estate farms, lead to better economic outcomes, fewer emissions, and similar water requirements. Nevertheless, to gain access to European markets, Malawi would need to reduce emissions from ethanol plants. We find that biofuels' economic and emissions outcomes are generally preferable to tobacco or soybeans. We conclude that the sustainability criteria encourage more sustainable biofuel production in countries like Malawi, but are perhaps overly biased against biofuels since other export crops raise similar concerns about food security and environmental impacts.

3.1 Introduction

Climate change, combined with population and economic growth, are placing tremendous pressure on natural resources. Managing these stresses requires a better understanding of the linkages between food, energy and water systems. Biofuels are a prime example of how advances in one system may come at the expense of others. Producing biofuels in developing countries could raise incomes and reduce dependence on imported fossil fuels (Msangi and Evans, 2013).

---

1 The article in this chapter is published as Schuenemann, F., Thurlow, J., and Zeller, M., 2017. Leveling the Field for Biofuels: Comparing the Economic and Environmental Impacts of Biofuel and Other Export Crops in Malawi. *Agricultural Economics*, 48: 301–315. doi:10.1111/agec.12335, and is used with permission of the authors. Previously, a similar version of the article in this chapter was also published online as a non-peer-reviewed IFPRI Discussion Paper 01500 (working paper) in 2016.

However, clearing lands for biofuel crops generates greenhouse gas (GHG) emissions (Fargione et al., 2008) and might worsen food insecurity by diverting resources from food production. Biofuels' greater use of water resources relative to fossil fuels is a further concern (Berndes, 2002).

Understanding biofuel's economic and environmental trade-offs is challenging, not least because evidence on the effects of biofuels is mixed. While the spike in global food prices in the late-2000s was partly attributed to global biofuels (Rosegrant et al., 2008), a more recent review by Zilberman et al. (2013) found no definite direction of impact. National studies also find that higher incomes from biofuels can offset higher food prices and *improve* food security (Ewing and Msangi, 2009; Arndt et al., 2010; Arndt et al., 2012; Negash and Swinnen, 2013). Finally, while Berndes (2002) projects that the water used by bioenergy crops will eventually equal that of existing crops, the conclusion of this and other studies is that there is enough water available *globally* to expand biofuel production (De Fraiture et al., 2008). Despite this mixed evidence, one conclusion that can be drawn from the literature is the need for more context-specific and integrated analysis.

Economic and environmental trade-offs also have major implications for biofuel policy. Many developing countries see biofuels as an export opportunity. But this hinges on gaining access to European Union (EU) markets, where preferential trade agreements enhance competitiveness and biofuel mandates ensure import demand. In response to concerns over biofuels, the EU introduced "sustainability criteria" that impose strict conditions for accessing biofuel markets (EC, 2010). By 2018, biofuels in the EU will be required to generate 60 percent fewer GHG emissions per liter than fossil fuels. Other criteria include avoiding excessive water use and food crop displacement. Governments in developing countries therefore need to know *in advance* how producing biofuels will affect emissions, food security and water use, and whether the EU's sustainability criteria preclude certain biofuel production arrangements.

Despite the need for integrated analysis, most biofuel studies are sector-specific or focus on specific outcomes, such as food production and prices, or land use change and GHG emissions (see, for example, Searchinger et al., 2008; Timilsina et al., 2012). This overlooks linkages between food, energy and water systems, and between biofuel industries and the rest of the economy. To address this limitation, we build on recent studies that use country-level computable general equilibrium (CGE) models to estimate the impacts of biofuel production on economic growth and poverty (Arndt et al., 2010, 2012) and GHG emissions (Thurlow et al., 2015). We extend this approach to include a more detailed treatment of the agricultural and natural resources required to produce biofuels vis-à-vis other crops. The CGE model is linked to biophysical models that estimate crop water requirements, and GHG emissions from direct and indirect land use change. The integrated modeling framework is applied to Malawi, where the government is debating whether to promote biofuels over other export crops. We compare sugarcane-ethanol production under different farming systems, including smallholder versus estate farms on irrigated versus rain-fed lands.

An important limitation of recent CGE-based studies is that they assumed that the "status quo" is the correct counterfactual for assessing biofuel impacts. These studies allowed new lands to be cleared for growing biofuel crops and this explained some of the resulting increases in national incomes. However, if cleared lands are not used for biofuels then they could be used for other crops. The correct counterfactual should assume that uncultivated lands do not remain idle in the absence of biofuels. This has important policy implications. The EU's sustainability criteria only apply to biofuels even though producing non-biofuel crops also has economic and environmental implications. The EU has therefore "raised the bar" on biofuel exports from developing countries, but it has also created an uneven "playing field". By comparing biofuels to other crops, our study can determine if biofuels from Malawi are of particular concern, or if the sustainability criteria are overly-biased against biofuels and so should either be relaxed or applied to other agricultural exports.

The chapter is structured as follows. Section 3.2 briefly describes the Malawian economy and the role of sugarcane and ethanol. Section 3.3 describes our integrated suite of models and Section 3.4 presents our simulation results. We conclude by summarizing our findings and discussing their implications for biofuel policy in Malawi, the EU, and elsewhere.

3.2 Biofuels in Malawi

3.2.1 Food, energy, and water systems

Agriculture accounts for a third of Malawi's gross domestic product (GDP) and four-fifths of employment. Most farmers are poor smallholders growing food crops for subsistence, although many also grow tobacco, which is Malawi's main export. Due to declining global tobacco demand, Malawi's government is searching for alternative export crops and biofuels is one of the options being considered (GOM, 2012). Malawi also imports its fossil fuels and so biofuels could help reduce severe foreign exchange constraints.

There are strong linkages between Malawi's food, energy, and water systems. A quarter of the country is covered by Lake Malawi and so irrigation potential is high and water scarcity should be a minor concern. However, most smallholders practice rain-fed farming and Malawi experiences frequent droughts causing substantial economic losses (Pauw et al., 2011). Irrigation infrastructure is unaffordable for most smallholders and only four percent of cropland is irrigated (SMEC, 2015). Malawi's electricity supply mainly comes from hydropower and so reductions in dam water levels could lead to electricity shortages. There is competition over scarce land resources. Malawi is the second most densely populated country in Africa and the average smallholder cultivates less than one hectare. Agricultural land expansion is therefore severely constrained and so any new export crop is expected to cause some displacement of existing crops on smallholder lands.

3.2.2 Sugarcane-ethanol

A biofuel export strategy in Malawi would start from an established base. Malawi has produced sugarcane since the 1960s. Today, two large estate farms grow 80 percent of the feedstock with the rest produced via smallholder outgrower schemes. Malawian sugarcane is almost entirely irrigated and, thanks to favorable agro-climatic conditions, achieves yields of around 100 metric tons per hectare, which is high by international standards.

Ethanol production from sugarcane started in the late-1980s. Malawi introduced a 10–20 percent petrol-ethanol blending mandate, but is still far from achieving this target.[2] In 2010, only 18 million liters of ethanol were produced compared to 360 million liters of imported petroleum. Ethanol prices are pegged to the petroleum prices, making locally-blended and imported petroleum equally expensive.

Malawi could export biofuels to the EU and the Southern African Development Community (SADC). Malawi has preferential access to EU markets through the "Everything but Arms Initiative" and is part of SADC's Free Trade Area. Foreign investors have shown interest in producing biofuels in Malawi (GOM, 2012). One constraint is the availability of lands suitable for sugarcane. Kassam et al. (2012) estimate that 14,000 hectares of *uncultivated* land is available

2 The typical blending ratio for petrol in Malawi is 10 percent ethanol and 90 percent petrol (Mitchell, 2011).

for rain-fed sugarcane. Meanwhile, Malawi's irrigation investment plan intends to grow around 50,000 hectares of *irrigated* sugarcane (SMEC, 2015). Realizing Malawi's full irrigation potential, which Watson (2011) estimates at 300,000 hectares, would require substantial investments. Land availability is therefore a major constraint to biofuel production in Malawi.

3.2.3 Ethanol production technologies

Malawi's export strategy intends to expand sugarcane production by 75,000 hectares (GOM, 2012). This is only two percent of total crop land and is therefore unlikely to have economy-wide implications. We simulate a more ambitious biofuel export strategy in order to more accurately gauge economy-wide impacts. However, the outcomes estimated in our analysis are roughly proportional to the scale of biofuel expansion.

Three broad biofuel options are available to Malawi. Table 3.1 shows the production technologies used to produce 1000 million liters ethanol per year, assuming a conversion ratio of 70 liters of ethanol per metric ton of feedstock. One option continues to grow sugarcane on estate farms, where irrigated farming systems achieve high yields of 108 tons per hectare. This would require 132,000 hectares of land. Another option would use smallholder outgrower schemes that are either irrigated or rain-fed. On average, irrigated smallholders obtain 99 tons per hectare, which is close to estate farm yields and therefore has similar land requirements (i.e., 144,000 hectares). Rain-fed smallholders only obtain 42 tons per hectare and so require 340,000 hectares of land. Given Malawi's land constraints, the choice of technology or farming system will greatly influence the extent of land clearing and/or crop displacement.

Table 3.1: Sugarcane-ethanol production technologies

	Input requirements per 1000 million liters of sugarcane-based ethanol		
	Irrigated estates	**Irrigated outgrowers**	**Rainfed outgrowers**
Liquid yield (liter/mt)	70.0	70.0	70.0
Feedstock required (1000 mt)	14,286	14,286	14,286
Land yield (mt/ha)	108.0	99.0	42.0
Land required (ha)	132,000	144,000	340,000
Workers employed (people)	49,271	53,669	100,634

	Input requirements per 1000 million liters of sugarcane-based ethanol		
	Irrigated estates	**Irrigated outgrowers**	**Rainfed outgrowers**
Feedstock	48,899	53,298	100,263
Processing	371	371	371
Labor yield (people/mil. liters)	49.3	53.7	100.6
Foreign capital requirements (units)	23,568	12,142	9,984
Feedstock	13,584	2,158	0
Processing	9,984	9,984	9,984
Capital yield (units/mil. liters)	23.6	12.1	10.0

Source: Own estimates using farm budget survey data (Herrmann & Grote, 2015) and processing cost estimates (Quintero et al., 2012).

Table 3.1's technologies were derived from various sources. Information on smallholder production is from a recent survey of sugarcane outgrowers (Herrmann and Grote, 2015). Estate farm technologies are extracted from Malawi's Annual Economic Survey (NSO, 2014). Ethanol processing costs are from a study of processing plants in Tanzania by Quintero et al. (2012), updated to include Malawian feedstock and labor costs. Malawi currently produces ethanol from molasses at two low-capacity processing plants. We assume that better processing technologies would be used to expanded biofuel production, and that the same technology would be used irrespective of the farming system supplying the feedstock.

Large-scale ethanol processing is not particularly labor-intensive (Table 3.1). Most jobs created in the biofuel industry are in feedstock cultivation. Estate farms are less labor-intensive than outgrower schemes and so employment and wage outcomes will vary by farming system. Total ethanol production costs amount to US$0.63 per liter, which is competitive with imported petroleum. Even with recent price fluctuations, petroleum prices in Malawi have not fallen below US$1 per liter. In global markets, Malawian ethanol is only profitable at a crude oil price of US$77 per barrel[3]. However, Malawi is exempt from EU

3 Malawian biofuels is profitable at current oil price projections of US$88 per barrel or higher from 2020 onwards (IEA, 2015). Including a shadow value for carbon (via global emissions trading) reduces Malawi's US$77 threshold oil price. For example,

ethanol tariffs (about US$0.19 per liter) and this makes Malawi's ethanol supply price comparable to those of Brazil (US$0.47) and the United States (US$0.46) (Arndt et al. 2012). The EU's biofuel mandate and favorable trade policies are prerequisites for Malawi's biofuel export strategy.

In the next section we develop an integrated modeling framework that incorporates the different biofuel technologies and evaluates how the choice of farming system influences the economic and environmental impacts of producing biofuels in Malawi.

3.3 Measuring economic and environmental impacts

CGE models are essential when evaluating large-scale interventions that are expected to have economy-wide implications. The Malawi CGE model is linked "top-down" to two natural resource models that measure environmental impacts: (i) a crop model that estimates crop water use; and (ii) a carbon accounting model that estimates GHG emissions from land use change.

3.3.1 Measuring economy-wide impacts

We use the recursive dynamic CGE model described in Diao and Thurlow (2012) to measure the economic impacts of producing biofuels.[4] Producers and consumers in the model maximize profits and utility and interact in factor and product markets. Production functions in each sector determine output levels and allow imperfect substitution between factors based on their relative prices. Composite factors are combined with intermediate inputs using fixed input-output relationships. There is also substitution between domestic, import and export markets, with the decision on how much to trade based on relative domestic and foreign prices (inclusive of taxes and transaction costs). Malawi is a small country and so world prices are fixed.

The model is calibrated to a 2010 social accounting matrix (SAM) (Pauw et al., 2015) that includes information on production technologies for 58 sectors. Labor is separated by three education levels and rural and urban areas. Crop land

Kossoy et al. (2015) estimate that a carbon price of US$100 per ton CO_2eq is needed to limit global warming to 2°C. When applied to our biofuel scenarios, the carbon savings relative to fossil fuels could make Malawi competitive at an oil price of US$54 per barrel.

4 The model's variables and equations are provided in Tables 3.A1 and 3.A2 in the appendix.

is separated into small, medium and large smallholders and large-scale estates.[5] Given Malawi's land and labor constraints during peak production periods, factors are assumed to be fully employed but mobile across sectors. Only capital is sector-specific.

The distributional impacts of biofuel production are captured by separating households into representative groups based on location (rural or urban), farm size (small, medium or large) and per capita consumption quintiles. Households can produce for their own consumption or engage in product markets. Households are the main owners of land, labor and capital, and incomes are used to either consume goods, pay taxes, or save. Consumption is determined by a linear expenditure system of demand with income elasticities estimated using Malawi's 2010/11 Integrated Household Survey (IHS3) (NSO, 2012). Changes in poverty rates are estimated following the micro-simulation approach in Arndt et al. (2012). Households in IHS3 are mapped to household groups in the CGE model. Proportional real consumption changes from the CGE model are passed down to households in the survey, where poverty rates are recalculated using the official poverty line.

Three "closure rules" maintain macroeconomic consistency. First, foreign capital inflows are fixed (beyond what is needed to expand biofuels production) and the exchange rate adjusts to equate supply and demand of foreign exchange. Secondly, private savings rates are fixed and so rising incomes lead to higher savings and investment. Thirdly, government tax rates and recurrent spending growth are fixed, and the recurrent deficit adjusts to balance total revenues and expenditures. The domestic price index is the model's numeraire.

The model is solved annually over the ten-year period, 2010–2020. Parameters are updated between periods based on long-term trends in factor supplies, total factor productivity, population, government spending, and foreign capital inflows. Capital stocks in each sector are updated each year to reflect depreciation and previous-period investment. Sectors with above-average profits receive a larger share of new capital stocks than their share of installed capital.

5 The model aggregates family-owned and rented crop land. As with owner-occupied dwellings in national accounts, land value-added is paid to smallholders assuming that they rent their lands from themselves. For more information on land in the SAM and CGE model, see Pauw et al. (2015) and Diao and Thurlow (2012).

3.3.2 Simulating biofuels production

Three new sugarcane sectors are added to the model based on the production technologies summarized in Table 3.1. Each sugarcane feedstock sector has its own ethanol processing sector. Production in the biofuel sectors is initially fixed at effectively zero. We then expand the amount of capital invested in a particular ethanol processing sector causing output to expand and drawing in factors and intermediate inputs. The latter includes sugarcane feedstock, which has its own input requirements. Ethanol production in each scenario is gradually increased until it reaches 1000 million liters per year by the end of the ten-year simulation period. Capital in the biofuel sectors is assumed to be foreign-owned, and so a "foreign capital" factor is inserted into the model that is only used in the biofuel sectors. All profits earned by this capital are repatriated and include repayment for irrigation equipment.

We assume that all new ethanol produced in Malawi is exported. In reality, some may be used to meet the domestic blending mandate. However, as discussed below, demand for imported petroleum in our baseline scenario reaches 587 million liters by 2020. Even if Malawi achieved its 20 percent blending mandate, this would only require 12 percent of the 1000 million liters of ethanol produced in the model in 2020. Local blending mandates do not alter the fact that most of the ethanol produced in our simulations would need to be exported. Fortunately, ethanol and petroleum are close substitutes and so there is little difference from a macro-accounting perspective between (i) exporting ethanol and using the foreign exchange to pay for imported fuels; and (ii) reducing fuel imports by redirecting ethanol to domestic markets and forgoing the foreign exchange earnings. This symmetry means that, even though we simulate biofuel exports, our results would be largely unchanged if some locally-produced biofuels are used to meet domestic mandates.

Two rounds of scenarios are run for each production technology shown in Table 3.1. The first scenarios assume that 132,000 hectares of new land are cleared and used to grow biofuel feedstock, and this increases total land supply in the model. This is exactly the amount of land needed by estates to produce the targeted level of ethanol and so there is no need to displace existing crops. In contrast, the smallholder scenarios require more than 132,000 hectares of land and so there is crop displacement. Imposing binding land constraints ensures that our results are not biased in favor of more land-intensive smallholder production options.

The second round of scenarios assumes that only 14,000 hectares of land can be cleared, which is consistent with the suitability assessment in Kassam et al.

(2012). Competition over scarce land resources and the level of crop displacement become more pronounced. Total land and labor supplies are fixed and so non-biofuel sectors may contract depending on their relative factor intensities. Table 3.2 compares the technologies of biofuel and other crops in Malawi. On average, existing crops generate lower GDP per hectare and worker than the new sugarcane crops. Reallocating resources to biofuels should therefore lead to an increase in average value-added per hectare and worker. These technology differences between crops largely determine the economic outcomes and crop displacement effects in our simulations.

Table 3.2: Biofuel and existing crop production technologies, 2010

	Production		Water use		Labor	Value-added or GDP per unit of input		
	Area (1000ha)	Yield (mt/ha)	Total (1000m³)	Intensity (m³/ha)	Intensity (people/ha)	Land (per ha)	Labor (per person)	Water (per m³)
Existing crops	4,179	2.7	11,030	2,639	0.30	300	1,008	114
Maize	1,696	2.0	4,146	2,444	0.30	350	1,155	143
Other cereals	202	1.0	702	3,473	0.41	349	845	101
Root crops	441	4.7	1,095	2,480	0.22	260	1,177	105
Pulses	705	0.7	1,625	2,305	0.12	143	1,189	62
Horticulture	496	3.5	1,633	3,293	0.66	364	551	110
Oilseeds	335	0.9	889	2,653	0.08	115	1,516	43
Export crops	304	9.9	941	3,094	0.36	504	1,394	163
New sugarcane feedstock								
Irrigated estates	0	108.0	0	10,212	0.37	918	2,483	90
Irrigated outgrowers	0	99.0	0	9,509	0.37	907	2,457	95
Rainfed outgrowers	0	42.0	0	5,057	0.29	430	1,457	85
New reference crops								
Tobacco	0	1.2	0	2,404	0.52	632	1,221	263
Soybeans	0	1.2	0	3,721	0.39	418	1,072	112

Source: Own estimates using production data from FAOSTAT; employment data from IHS3 (NSO, 2012); value-added data from the 2010 social accounting matrix (Pauw et al., 2015); and estimated water use from the process-based crop models (see Section 3.3).

3.3.3 Estimating crop water use

Results from the CGE model are passed down to crop models that calculate crop water use. Sugarcane is a relatively water-intensive crop and so producing biofuels is expected to increase consumptive water use. Irrigation further increases water use since some water is lost due to inefficient irrigation management. We adopt a "yield response to water" approach (Doorenbos and Kassam, 1979), as reflected in the equation below:

$$\left(1-\frac{Y_a}{Y_m}\right)=K_y\left(1-\frac{ET_a}{ET_m}\right)$$

where Y_a and Y_m are actual and maximum potential yield (tons per hectare); K_y are crop-specific yield response factors; and ET_a and ET_m are actual and maximum potential evapotranspiration (millimeters per day). The model assumes a linear relationship between relative yield declines and relative water deficits, where the latter is the ratio of actual to potential evapotranspiration. The strength of this relationship depends on crop-specific yield response factors.

Potential evapotranspiration ET_m is the water a plant would use if water were always available. Reference evapotranspiration for a hypothetical grass crop was derived using the Penman-Monteith equation and data from 20 Malawian weather stations for the period 1983–2005. Potential evapotranspiration was calculated by multiplying reference evapotranspiration by a coefficient k_c, which captures crops' unique physiological properties. Coefficients for Malawi are from Allen et al. (1998) and Rosegrant et al. (2012). Actual evapotranspiration ET_a for rain-fed and irrigated crops was calculated using daily soil water balances derived from precipitation data, and soil data from the Africa Soil Profiles Database. Irrigation water use is computed so as to reduce irrigation frequency while avoiding crop water stress. The models estimate *net* irrigation requirements – they exclude water lost during the irrigation process.

We estimated crop water use during an average weather year in Malawi. Average potential evapotranspiration and potential yields were compared to observed yields in the CGE model to back-calculate actual evapotranspiration or water use. Table 3.2 summarizes our results and confirms that sugarcane is particularly water-intensive.[6] Irrigated estate sugarcane, for example, consumes four times more water per hectare than the average for existing crops. These numbers are

6 These are approximate water use estimates since our linear crop model may not capture water-yield responses during extreme weather events.

used to calculate how the water intensity of agriculture in Malawi changes following a simulated increase in biofuel production in the CGE model.

3.3.4 Measuring greenhouse gas emissions

Land use change in the CGE model affects both crop water requirements and GHG emissions. The "sustainability criteria" requires GHG emissions from biofuels to be 60 percent lower than fossil fuel emissions. The default average life cycle GHG emission of petroleum is 2.92 kilograms of carbon dioxide equivalents per liter (kgCO_2eq/l). If Malawi wants to export to EU markets then the maximum permissible emissions from ethanol production is 1.17 kgCO_2eq/l.[7] However, Dunkelberg et al. (2013) estimate that emissions from current ethanol-molasses production in Malawi is 4.04 kgCO2eq/l. Most of these emissions come from unsustainable handling of waste products and coal heating in ethanol processing. The authors estimate that if energy were derived from waste products, which is similar to the assumption in Quintero et al. (2012), then emissions drop to 2.05 kgCO2eq/l, of which 1.15 kgCO2eq/l is from processing. We use this estimate of processing emissions in our scenarios, and then add emissions from land clearing and feedstock production.

Feedstock emissions are estimated using a model called "Ex-ante Carbon-balance Tool" (EX-ACT) (Bernoux et al., 2011). This is a land-based accounting tool that calculates the carbon balance from GHG emissions and carbon sequestration in the soil following changes in land use and management. Soil sequestration values for Malawi come from the World Bank's Soil Carbon Sequestration Geodatabase, and the model is calibrated to the tropical conditions and soil types prevalent in Malawi's southern and central regions where sugarcane production is likely to occur.

Land clearing for sugarcane in the CGE model leads to direct land use change and the displacement of other crops.[8] We assume that grasslands are cleared since deforestation generates emissions that far exceed EU thresholds. Grassland conversion emits 12.9 tCO_2eq/ha, half of which are once-off emissions when lands

7 Emissions from ethanol fuel use are not included since these are set to zero in EU calculations. These are a large part of petrol emissions and should ideally be included in a complete life cycle analysis of ethanol.

8 We implicitly capture emissions from indirect land use change (ILUC) in the model, since emissions from land clearing are the same regardless of whether cleared lands are used for biofuel or other crops.

are first cleared. When sugarcane displaces existing crops, then the net emissions depend on the inputs used to grow each crop as well as the crops' soil organic carbon sequestration (SOC) potential. Two reference crops are used when estimating changes in net emissions. For displaced food crops, we use the SOC value of maize (0.617 tC/ha/yr) since this is Malawi's main staple crop. For export crops, we use the SOC value of soybeans (0.839 tC/ha/yr) since this is one of the most affected crops in our simulations. Irrigation is more input-intensive and so generates more emissions. Sugarcane itself, however, has heavier biomass and so is a carbon sink relative to the reference crops, with a SOC value of 1.220 tC/ha/yr. This means that, without land clearing, switching from existing crops to sugarcane leads to lower net emissions.

3.4 Results

3.4.1 Baseline scenario

The CGE model's baseline scenario tracks recent trends in population and economic growth. Labor and land supplies grow at 2.0 and 1.7 percent per year, respectively, and total factor productivity grows at 2.7 percent per year. This generates annual GDP growth of 4.7 percent (see Table 3.3) and this is fairly evenly distributed across sectors. Note that the baseline scenario is only of marginal interest for our analysis, since it merely provides a reference for measuring the impacts of expanding biofuel production. In discussing the impacts of biofuel crops, we first compare biofuels to this "status quo" baseline scenario (as in previous studies), and then later to alternative reference scenarios that allow for the expansion of other export crops.

Table 3.3: Production and price impacts

	Initial share or value, 2010	Baseline growth rate or total change (%)	Deviation from final year baseline value (%)					
			Biofuel scenarios with land expansion			Biofuel scenarios with land constraints		
			Irrigated estates	Irrigated outgrowers	Rainfed outgrowers	Irrigated estates	Irrigated outgrowers	Rainfed outgrowers
Total GDP growth (%)	100.0	4.7	2.0	1.7	0.9	1.8	1.4	0.7
Agriculture	32.3	4.6	2.9	3.1	1.5	1.8	2.1	0.4
Food crops	16.6	4.5	1.9	2.9	-0.4	0.0	0.9	-2.7
Export crops	3.1	4.5	22.7	18.4	20.2	21.7	17.5	19.5
of which non-biofuels	3.1	4.5	-18.3	-25.8	-29.0	-19.3	-26.7	-29.8
Other agriculture	11.2	4.8	-0.8	-0.5	-0.5	-0.7	-0.3	-0.4
Industry	16.5	5.6	1.2	-1.0	-1.2	1.7	-0.5	-0.7
of which ethanol	0.0	0.0	∞	∞	∞	∞	∞	∞
of which electricity	0.8	4.2	23.7	20.4	18.1	23.8	20.6	18.3
Services	51.2	4.5	1.7	1.6	1.3	1.8	1.7	1.3
Change in price indices (%)								
Real exchange rate	1.0	6.0	-2.7	-3.4	-3.2	-2.4	-3.2	-2.8
Real food prices	1.0	4.0	-0.5	-0.4	0.2	-0.3	-0.1	0.6
Total crop land (1000ha)	4,233	777	132	132	132	14	14	14
Food crops	3,357	841	72	104	-43	-16	13	-140
Existing export crops	639	-71	-72	-116	-165	-102	-143	-186
Feedstock crops	0	0	132	144	340	132	144	340

Source: Results from the Malawi CGE model.

Notes: Biofuels processing grows from a zero base and so growth is infinite.

3.4.2 Producing biofuels on estate farms

We initially focus on the first three biofuel scenarios in which 132,000 hectares of uncultivated lands are cleared for sugarcane-ethanol. The third column in Table 3.3 reports final year deviations from baseline for the Irrigated Estate scenario. There is no direct crop displacement in this scenario since newly cleared lands exactly equal the amount of land required to grow feedstock on estate farms (see Table 3.1). There is, however, indirect land use change. Lands are reallocated from existing export crops (e.g., tobacco) to food crops (e.g., maize). This is driven by biofuel exports, which grow rapidly and cause the real exchange rate to appreciate, thereby reducing the competitiveness of non-biofuel export crops in foreign markets. To some extent, Malawi exchanges one export crop for another. However, since value-added per hectare for sugarcane is higher than it is for existing export crops (see Table 3.2), switching to biofuels leads to higher agricultural GDP. The clearing of new lands also increases the supply of productive resources. Higher incomes and an appreciated exchange rate lead to more land allocated to food production and lower real food prices. Unlike in the Arndt et al. (2010) study for Mozambique – a country with few non-biofuel export crops – we find that biofuels production in Malawi might eventually lead to improved food availability.

Table 3.4 reports impacts on labor and households. Bringing newly cleared lands into production increases demand for labor on estate farms. However, food crops and estate farms are less labor-intensive than displaced export crops (see Table 3.2), causing agriculture's overall labor share to decline. Rural wages still increase due to higher agricultural GDP, but the gains in urban wages are larger. Urban workers benefit from rising labor demand in ethanol processing sectors, but this is more than offset by falling employment in sectors that process existing crops (e.g., tobacco curing). The increase in non-farm employment and urban wages mainly comes from workers migrating to the trade and business sectors, which benefit from higher incomes and greater demand for non-traded services. Growth in industrial GDP is driven by increased electricity generation following the expansion of ethanol processing and irrigated estates, both of which are more energy-intensive than the manufacturing sectors and export crops that they displace.

Table 3.4: Labor and household impacts

	Initial value or share, 2010	Baseline growth rate or total change (%)	Deviation from final year baseline value (%)					
			Biofuel scenarios with land expansion			Biofuel scenarios with land constraints		
			Irrigated estates	Irrigated outgrowers	Rainfed outgrowers	Irrigated estates	Irrigated outgrowers	Rainfed outgrowers
Agriculture labor share (%)	63.5	64.2	-0.9	1.0	4.4	-1.4	0.6	4.0
Real wage (%)	3725.7	2.6	1.1	1.3	1.0	0.9	1.0	0.7
Rural workers	3,617	2.7	0.7	1.0	1.0	0.4	0.7	0.7
Urban workers	3,835	2.6	1.5	1.6	0.9	1.3	1.4	0.7
Household welfare (%)	425.7	1.6	0.7	1.3	0.6	0.4	1.0	0.2
Farm households	330	1.6	0.2	1.6	0.8	-0.2	1.1	0.3
Non-farm households	1,019	1.5	1.5	0.8	0.3	1.5	0.7	0.2
Poverty headcount rate (%)	51.0	28.7	-0.1	-2.4	-0.9	0.8	-1.4	0.2
Farm households	55.9	32.1	-0.1	-2.4	-1.0	0.9	-1.4	0.1
Non-farm households	20.3	7.8	-2.5	-1.6	0.9	-1.3	0.0	2.1

Source: Results from the Malawi CGE and microsimulation models.

Notes: Welfare is measured using real consumption expenditure, the initial value is average per capita US$ expenditure. Poverty headcount rate is the share of the population with per capita expenditures below the national poverty line.

Ultimately, national GDP is two percent higher in the Irrigated Estate scenario than in the baseline. This positive growth-effect from biofuels is driven by (i) an increase in the level of productive resources in the economy, i.e., from newly cleared lands and additional foreign capital; (ii) higher value-added per worker and per hectare of cropland in the biofuel sectors; and (iii) positive spillover or growth linkage effects, e.g., incomes from biofuels generating demand for all goods and services. Realizing these medium-term economic gains will impose adjustment costs on the economy, particularly on producers of existing crops and workers in downstream agro-processing. It may also require additional public investments in the electricity sector, the cost of which is only partially internalized in our model.

The household welfare and distributional effects of producing biofuels on Estate farms are less promising than the macroeconomic results would suggest. Household welfare does not increase by as much as GDP, because the profits from biofuel production are repatriated. Smallholders do not benefit by as much as non-farm households. This is because smallholders previously produced export crops, like tobacco, but these were displaced by sugarcane grown in estates. Smallholders find themselves growing more food crops and relying more on wages from estate farms, both of which generate less income than the displaced export crops (see Table 3.2). In contrast, non-farm households benefit from lower food prices, higher urban wages, and cheaper imports. The reduction in the urban poverty rate (by 2.5 percentage points) is therefore larger than the reduction in rural poverty (by only 0.1 percentage points). Producing biofuels on estate farms reduces poverty and improves food availability, but the benefits for the rural poor are fairly modest.

3.4.3 Using outgrower schemes

Smallholders achieve lower yields and require more land to meet the biofuel production target (see Table 3.1). The fourth column of Table 3.3 reports results for the Irrigated Outgrower scenario. Producing sugarcane feedstock now requires 144,000 hectares, but we still only allow 132,000 hectares of new lands to be cleared. This means that biofuels directly displace existing crops. As in the previous scenario, the real appreciation caused by biofuel exports directs all of the displacement onto existing export crops. In fact, the land allocated to food crops increases. The appreciation is larger in the Irrigated Estate scenario, because more of the on-farm profits from growing sugarcane via outgrower schemes remains with smallholders rather than being repatriated to foreign investors. Higher smallholder incomes also generate greater demand for products that smallholders consume intensively, such as food. The reallocation of land from existing export crops to food crops is therefore larger in this scenario. Higher demand for food also means that real food prices fall by less than in the previous scenario even though the increase in food production is now larger.

The large decline in existing export crops in this scenario leads to larger job losses in downstream processing. This more than offsets the expansion in ethanol processing, leading to lower manufacturing GDP and employment. However, smallholder farming is more labor-intensive than estate farming (see Table 3.1) and so using outgrower schemes increases agriculture's labor requirements. Overall, agriculture's employment share rises in the Irrigated Outgrower scenario and is matched by higher agricultural GDP growth (see Table 3.4). Gains in national GDP are smaller in this scenario due to declining industrial GDP. Faster agricultural growth is coupled with larger improvements in household welfare. Outgrower schemes mean that more benefits from producing biofuels accrue to smallholders. It is rural rather than urban households that now experience the largest gains in welfare and poverty reduction.

3.4.4 Relying on rain-fed production

In the Irrigated Outgrower scenario above we assumed that foreign investors provided irrigation infrastructure and that smallholders in the outgrower scheme repaid investors over a ten year period. This explains why irrigated smallholder farms require foreign capital (see Table 3.1). We now consider the implications of producing sugarcane using smallholder farmers who do not have access to irrigation. Since there is no irrigation, smallholders no longer use foreign capital and do not have to repay investors. However, without irrigation, smallholders achieve much lower yields and require 340,000 hectares of land in order to achieve the ethanol production target. Again, we assume that only 132,000 hectares of new lands can be cleared.

Results are shown in the fifth column of Table 3.3. The level of crop displacement caused by biofuels is sufficiently large such that there is now a decline in the lands allocated to *both* existing export and food crops, although impacts on the former are still more pronounced. Declining food production leads to higher real food prices. There is still a positive effect on national GDP, but this is now smaller than before because land and labor productivity gains are more modest. For example, value-added per hectare of sugarcane in the Rainfed Outgrower scenario is US$430, which is much lower than the US$907 in the Irrigated Outgrower scenario (see Table 3.2). This explains the smaller increase in agricultural

GDP. Again, industrial GDP falls slightly because of a contraction in downstream agro-processing.

Rainfed sugarcane has a higher labor-to-land ratio than existing crops (see Table 3.2) and so reallocating land leads to a higher average labor intensity for agriculture as a whole. Agriculture's share of employment increases substantially by 4.4 percentage points in the Rainfed Outgrower scenario. Higher agricultural labor demand helps maintain rural wage growth despite slower agricultural GDP growth. Slower non-agricultural growth, on the other hand, means slower urban wage growth. Household welfare still improves for both rural and urban households, but the gains are smaller than under the Irrigated Smallholder scenario. Urban poverty rates rise because of higher real food prices.

The results from the first three scenarios suggest that biofuels can generate economic growth and reduce poverty. This can be achieved without jeopardizing food security only if feedstock is grown on irrigated lands. Finally, the choice between irrigated estate farms or irrigated outgrower schemes involves a clear trade-off between maximizing national growth or poverty reduction.

3.4.5 Environmental impacts and trade-offs

Table 3.5 reports the estimated GHG emissions and crop water use associated with the three biofuel scenarios discussed above. The final line in the table shows the amount of water used to grow sugarcane per liter of ethanol produced. Water use is much higher under irrigation. Even assuming a high irrigation efficiency rate of 50 percent, irrigated smallholder cultivation uses almost twice as much water as rainfed sugarcane (i.e., 3,387 versus 1,720 liters). Rainfed agriculture is most efficient in its water use, but it achieves lower yields and uses more land.

Table 3.5: Emissions and water use

	Biofuel scenarios with land expansion			Biofuel scenarios with land constraints		
	Irrigated estates	Irrigated outgrowers	Rainfed outgrowers	Irrigated estates	Irrigated outgrowers	Rainfed outgrowers
GHG emissions embodied within ethanol ($kgCO_2eq/l/yr$)						
After 1 year	2.61	2.62	2.31	1.42	1.46	1.00
After 10 years	1.81	1.82	1.52	1.34	1.37	0.91
After 20 years	1.77	1.78	1.47	1.33	1.37	0.91
Decomposition of GHG emissions after first year ($kgCO_2eq/l/yr$)						
Clearing lands	1.70	1.70	1.70	0.18	0.18	0.18
Displacing crops	0.00	0.04	0.57	0.34	0.40	0.78
Growing feedstock	-0.24	-0.27	-1.11	-0.24	-0.27	-1.11
Processing feedstock	1.15	1.15	1.15	1.15	1.15	1.15
Crop water use (mil. m^3)						
Total crop water use	1,435	1,435	1,423	1,407	1,407	1,395
of which feedstock	135	137	172	135	137	172
Change from baseline	137	136	125	109	109	97
Crop water embodied within ethanol (liters/liter)	3,198	3,387	1,720	3,198	3,387	1,720

Source: Results from the Malawi CGE and crop models.

Notes: GHG emissions are measured by tons of CO_2 equivalent per liter of ethanol per year.

Our estimated water requirements are fairly high. Gerbens-Leenes and Hoekstra (2009) estimated that Brazil, as the world's largest ethanol producer, uses 2,500 liters of water per liter of ethanol. These authors included polluted "grey water" from fertilizers, whereas we consider only crop and irrigation water. Malawi's actual water use may therefore be higher than our estimates. In total, we find that 2 billion cubic meters of water is needed per year to produce 1000 million liters of ethanol. This appears to be small relative to the 8,400 billion cubic meters of water in Lake Malawi. However, local impacts on small watersheds can be significant. The effects on water levels in Lake Malawi and the Shire River would need to be determined by hydrological water basin models.

Table 3.5 reports changes in consumptive water used by crops in Malawi relative to the baseline. This includes total crop evapotranspiration, but not the water lost through inefficient irrigation. Rainfed sugarcane uses more water than irrigated sugarcane, i.e., 172 and 137 million cubic meters, respectively. This is because, even though rainfed sugarcane uses less water per hectare, it also uses more land. Rainfed sugarcane therefore displaces more crops with lower water needs (see Table 3.2), thus driving up total evapotranspiration. Overall, sugarcane-ethanol expansion in our simulations increases the water intensity of Malawian agriculture by almost ten percent irrespective of which biofuel production technology is used and despite the fact that land area increases by only 2.6 percent.

We now consider emissions from land use change. In the three biofuel scenarios discussed above we assumed that 132,000 hectares of land are cleared to grow sugarcane. Table 3.5 indicates that most of the emissions per liter of ethanol are from the once-off clearing of grasslands. Over time, these emissions are spread over a larger volume of ethanol until eventually the emissions per liter are essentially only those from cultivating sugarcane, such as the fossil fuels used for fertilizer, irrigation and transport. The reported emissions are therefore higher in the first year of ethanol production and lower after ten years.

Processing ethanol in Malawi generates 1.15 tCO_2eq/l, which is close to the EU's 2018 threshold of 1.17 $kgCO_2eq/l$. Emissions from growing feedstock therefore have to be extremely low in order for Malawi to export to EU markets. Our results indicate that clearing 132,000 hectares of grassland generates 1.70 $kgCO_2eq/l$ in the first year of ethanol production. Although sugarcane is a carbon sink, it cannot offset the emissions from land clearing. Even after 20 years, emissions in the Rainfed Outgrower scenario are 1.47 tCO_2eq/l. If sugarcane is grown on irrigated lands with inputs that directly or indirectly use fossil fuels then emissions are higher at 1.78 tCO_2eq/l.

The potential for Malawian biofuels to help mitigate climate change is hampered by the high carbon debt from land clearing and the emissions from ethanol processing. The latter should be kept in mind when building new processing plants. Dunkelberg et al. (2014) find that if ethanol plants in Malawi switch from coal to energy produced using crop residues, then the processing emissions become almost negligible. While this would require investments in more sophisticated technologies, it would improve Malawi's chances of meeting the EU's targets.

3.4.6 Imposing stricter land constraints

The previous scenarios assumed that 132,000 hectares of land are cleared for sugarcane. This may not accurately reflect Malawi's severe land constraints and so might exaggerate growth and welfare gains. This section assumes that only 14,000 hectares of land suitable for sugarcane is cleared. This is consistent with land suitability estimates from Kassam et al. (2012), but is much more conservative than SMEC (2015) or Watson (2011). The final three columns in Table 3.3 report results for biofuel production under land constraints. Each scenario should be compared to the corresponding scenario that allowed for more land expansion.

We still produce the same targeted level of ethanol as in earlier scenarios. However, stricter land constraints mean that there is greater displacement of existing crops, particularly for export crops. The land allocated to food crops now declines in the Irrigated Estate scenario and is much smaller in the Irrigated Outgrower scenario. Falling food production means higher real food prices. This is clearest in the Rainfed Outgrower scenario, where almost all export crops are displaced and there is a large reduction in lands for food crops.

Average value-added per hectare still rises in the biofuel scenarios due to sugarcane's higher land productivity relative to existing crops. However, the gains in agricultural GDP are smaller because less new land is added to productive resources. Less land also means smaller increases in demand for farm labor and so agriculture's share of employment does not increase by as much (see Table 3.4). Slower growth in labor demand means smaller wage increases and welfare improvements. The Irrigated Outgrower scenario generates the largest welfare gains for poorer households. In contrast, poverty in the Rainfed Outgrower scenario actually increases due to higher food prices and smaller land productivity and wage gains. The economic trade-offs between production technologies are starker with stricter land constraints.

As expected, emissions per liter of ethanol are much lower when there is less land clearing (see Table 3.5). After ten years, emissions in the Rainfed

Outgrower scenario are only 0.91 tCO_2eq/l, which is well below the EU's target of 1.17 tCO_2eq/l. However, rainfed production is not an attractive option given its adverse effects on food production and poverty. The Irrigated Outgrower scenario is preferable from a development perspective, but, even after ten years, its emissions per liter of ethanol exceed the EU target.

Feedstock water use per liter of ethanol is unchanged in the land constrained scenarios. However, total crop water use declines slightly because of the greater displacement of existing crops. If incremental water use from producing biofuels is used to measure biofuels' water content (i.e., if we deduct displaced crop water use), then additional water use resulting from producing biofuels is 20 percent lower than in the previous scenarios (e.g., 109/136=0.80 for the Irrigated Outgrower scenarios). Yet, even with this more lenient measurement, ethanol's water use in Malawi still exceeds that in Brazil. Overall, our analysis suggests that *both* development and environmental objectives could be achieved if Malawi reduces its emissions from ethanol processing; and if no quantitative restrictions on water use are added to the EU's sustainability criteria.

3.4.7 Biofuels versus other export crops

So far we have followed the approach of previous studies by comparing biofuel production to a "status quo" baseline. Yet Malawi's export strategy (GOM, 2012) suggests that if croplands are not used to grow sugarcane for ethanol, then they might be used to grow other export crops. A more appropriate counterfactual should consider these opportunity costs. We consider three alternative export crops: tobacco, soybeans, and sugarcane grown for refined sugar production (as opposed to ethanol production). Tobacco is a well-established smallholder crop with strong downstream linkages to agro-processing (i.e., tobacco curing). Soybeans is a relatively new crop that is identified alongside biofuels in the export strategy. Finally, using sugarcane feedstock to produce refined sugar rather than ethanol is an important option given fluctuations in global oil and ethanol prices.

In our new counterfactual "cash crop" scenarios, we assume that the additional tobacco and soybean production and downstream processing make use of Malawi's existing technologies (as captured in the official 2010 SAM). The only difference between sugarcane grown for ethanol and refined sugar is in how the feedstock is processed – the actual feedstock is grown using the same smallholder outgrower schemes (see Table 3.2). As with biofuels, tobacco, soybean and sugar refining are entirely financed by foreign capital and all profits are repatriated. To ensure that our scenarios are comparable, we simulate the same 144,000 hectare expansion in crop land devoted to each alternative cash crop and only permit 14,000 hectares of new lands to be cleared. We compare the new cash crop reference scenarios to the earlier Irrigated Outgrower scenario that produced ethanol.

Table 3.6 reports results for the Tobacco, Soybean and Sugar scenarios alongside results from the Irrigated Outgrower ethanol scenario. Agricultural exports increase in all four scenarios, but food production falls in the Tobacco and Soybean scenarios (relative to the baseline). Tobacco and soybeans have high labor-to-land ratios (see Table 3.2) and so expanding their cropland area draws labor away from food crops. A larger exchange rate appreciation in the Sugar scenario leads to greater displacement of export crops and production of food crops. Agriculture's share of total employment in the Tobacco and Soybean scenarios increases by 5.6 and 4.6 percentage points, respectively, compared to a one percentage point increase in the Irrigated Outgrower scenario and a slight reduction in the Sugar scenario. Both land and labor displacement are important in determining impacts on food production.

Table 3.6: Comparing biofuels to alternative cash crops

	Deviation from final year baseline value (%)			
	Ethanol (irrigated outgrowers)	**Tobacco (rainfed outgrowers)**	**Soybeans (rainfed outgrowers)**	**Sugar (irrigated outgrower)**
Total crop land (1000ha)	5,024	5,024	5,024	5,024
of which cleared lands	14	14	14	14
Total GDP growth (%)	1.4	0.4	0.0	1.8
Agriculture	2.1	1.2	0.5	2.1
Food crops	0.9	-0.7	-1.3	1.5
Export crops	17.5	17.1	12.2	14.0
Industry	-0.5	0.7	0.7	3.9
Services	1.7	-0.2	-0.5	0.8
Change in price indices (%)				
Real exchange rate	-3.2	-1.0	-0.3	-3.6
Real food prices	-0.1	0.5	0.5	0.1
Household welfare (%)	1.0	0.2	-0.1	1.3
Farm households	1.1	0.4	0.1	1.9
Non-farm households	0.7	-0.4	-0.4	0.2

	Deviation from final year baseline value (%)			
	Ethanol (irrigated outgrowers)	Tobacco (rainfed outgrowers)	Soybeans (rainfed outgrowers)	Sugar (irrigated outgrower)
Poverty headcount rate (%)	-1.4	-1.1	-0.4	-3.7
Farm households	-1.4	-1.2	-0.4	-3.7
Non-farm households	0.0	0.3	0.3	-2.3
Total crop water use (mil. m^3)	1,407	1,333	1,377	1,407
Emissions from feedstock production (tCO_2eq/yr)				
Per ha after 10 years	1.6	1.7	0.7	1.6
Per additional US$ GDP	1.6	6.4	34.0	1.6

Source: Results from the Malawi CGE and microsimulation models.

Notes: Welfare is measured using real consumption expenditure, the initial value is average per capita US$ expenditure. Poverty headcount rate is the share of the population with per capita expenditures below the national poverty line.

Tobacco and soybeans generate less value-added per hectare than sugarcane and so agricultural GDP gains are smaller. All four crops create downstream jobs, but tobacco and soybean processing and sugar refining are more labor-intensive. Agriculture's rising labor-intensity along with the creation of more industrial jobs means that fewer workers migrate to the service sectors. At the same time the high labor intensity of sugar refining leads to much larger industrial growth than in the other scenarios. Total GDP is unchanged in the Soybean scenario because of this crop's low land productivity. Agricultural gains are exactly offset by nonagricultural losses. Tobacco production leads to an increase in total GDP, but these gains are smaller than in the Irrigated Outgrower scenario. Overall, using sugarcane for refined sugar leads to a larger increase in total GDP growth than using it for ethanol. This underscores the importance of including opportunity costs in the counterfactual scenario.

Changes in household welfare mirror the changes in total GDP. Farm household poverty declines in both the Tobacco and Soybean scenarios, reflecting the importance of these crops for poorer smallholders in Malawi. However, total welfare gains and poverty reduction are much larger in the two sugarcane-based scenarios. This is consistent with Herrmann and Grote (2015), who found that sugarcane outgrower schemes reduce poverty amongst smallholder farmers in

Malawi. Even though biofuels production leads to better economic outcomes than either tobacco and soybeans, we find that the production of refined sugar is even more beneficial in terms of both welfare gains and poverty reduction.

Finally, we compare environmental impacts. Total crop water use in Malawi increases in the Tobacco and Soybean scenarios because these crops use more water per hectare than the crops they displace. Ethanol and refined sugar production are, however, much more water-intensive than tobacco or soybeans. This justifies the concerns about the pressure that biofuels place on scarce water resources. Importantly, emissions per hectare of sugarcane are lower than emissions from tobacco. Sugarcane's emissions per dollar of crop GDP is also lower than either tobacco or soybeans.[9] All alternative cash crops generate positive emissions, but sugarcane's heavy biomass means that it is a larger carbon sink than soybeans and tobacco (this is despite sugarcane being irrigated and using more fertilizers).

3.5 Conclusion

We developed an integrated modeling framework that jointly assesses the economic and environmental impacts of producing biofuels in Malawi. We find that sugarcane production on large-scale estate farms has the largest positive effect on economic growth. However, irrigated outgrower schemes are more effective at reducing poverty. Smallholders in Malawi that use irrigation achieve sugarcane yields that are similar to those on estate farms, and so the level of GHG emissions per liter of ethanol is similar across these two irrigated farming systems. Reliance on rainfed cropping systems leads to far less favorable food security and poverty outcomes although it does generate lower GHG emissions. There are clear trade-offs between each farming system. Nevertheless, we conclude that irrigated smallholder outgrower schemes operating on existing crop lands is the preferred means of producing biofuels in Malawi.

More generally, we conclude that concerns about the impacts of biofuels on climate change and food security are warranted in the case of Malawi. Producing biofuels in Malawi increases crop water use and GHG emissions. However, similar concerns can also be raised about other export crops. Our analysis for Malawi suggests that growing sugarcane generally leads to better economic outcomes and fewer GHG emissions than tobacco or soybeans – two crops that feature

9 Emissions are from crops cultivation only and exclude possible processing emissions. Estimates suggest that emissions from tobacco curing are higher than those of ethanol processing. No estimates were available for soybeans.

prominently alongside biofuels in Malawi's new export strategy. The EU's sustainability criteria are correct in seeking to "raise the bar" on the environmental standards that must be met by biofuel producers. However, our study also suggests that these criteria are perhaps overly-biased against biofuels. A "level playing field" should impose similar economic and environmental standards on all agricultural exports from developing countries like Malawi.

References

Allen, R.G., L.S. Pereira, D. Raes, Smith, M., 1998. Crop Evapotranspiration – Guidelines for Computing Crop Water Requirements. *Irrigation and Drainage paper 56*, United Nations Food and Agriculture Organization, Rome, Italy.

Arndt, C., Benfica, R., Tarv, F., Thurlow, J., Uaiene, R., 2010. Biofuels, poverty, and growth: a computable general equilibrium analysis of Mozambique. *Environmental and Development Economics*, 15, 81–105.

Arndt, C., Pauw, K., Thurlow, J., 2012. Biofuels and economic development: A computable general equilibrium analysis for Tanzania. *Energy Economics*, 34(6), 1922–1930.

Berndes, G., 2002. Bioenergy and water: The implications of large-scale bioenergy production for water use and supply. *Global Environmental Change*, 12, 253–271.

Bernoux, M., Tinlot, M., Bockel, L., Branca, G., Gentien, A., 2011. *EX-ante carbon-balance tool (EX-ACT): technical guidelines for Version 3, Easypol module 101*. Food and Agriculture Organization of the United Nations, Rome, Italy.

De Fraiture, C., Giordano, M., Liao, Y., 2008. Biofuels and implications for agricultural water use: blue impacts of green energy. *Water Policy*, 10(S1), 67–81.

Diao, X., Thurlow, J., 2012. A Recursive Dynamic Computable General Equilibrium Model, in X. Diao, J. Thurlow, S. Benin and S. Fan, eds., *Strategies and Priorities for African Agriculture: Economywide Perspectives from Country Studies*. International Food Policy Research Institute (IFPRI), Washington, DC.

Doorenbos, J., Kassam, A.H., 1979. Yield Response to water. *Irrigation and Drainage paper 33*, United Nations Food and Agriculture Organization, Rome, Italy.

Dunkelberg, E., Finkbeiner, M., Hirschl, B., 2013. Sugarcane ethanol production in Malawi: Measures to optimize the carbon footprint and to avoid indirect emissions. *Biomass and Bioenergy*, 71, 37–45.

EC (European Commission), 2010. Communication from the Commission on the practical implementation of the EU biofuels and bioliquids sustainability scheme and on counting rules for biofuels. *Official Journal of the European Union*, C160, 8–16.

Ewing, M., Msangi, S., 2009. Biofuels production in developing countries: assessing tradeoffs in welfare and food security. *Environmental Science and Policy*, 12, 520–528.

Fargione, F., Hill, J., Tilman, D., Polasky, S., Hawthorne, P., 2008. Land clearing and the biofuel carbon debt. *Science*, 319, 1235–1238.

Gerbens-Leenes, P.W., Hoekstra, A.Y., 2009. The Water Footprint of Sweeteners and Bio-Ethanol from Sugar Cane, Sugar Beet and Maize. Value of Water Research Report Series No. 38, UNESCO-IHE Institute for Water Education, Delft, the Netherlands.

GOM (Government of Malawi), 2012. National Export Strategy (NES) 2013–2018, Volume 2, Annexes 1–5, Lilongwe, Malawi.

Herrmann, R., Grote, U., 2015. Large-scale Agro-Industrial Investments and Rural Poverty: Evidence from Sugarcane in Malawi. *Journal of African Economies*, 24(5), 1–32.

IEA (International Energy Agency), 2015. World Energy Outlook 2015. Paris, France.

Kassam, A., Lutaladio, N., Friedrich, T., Kueneman, E., Salvatore, M., Bloise, M., Tschirley, J., 2012. Natural Resource Assessment for Crop and Land Suitability: An application for selected bioenergy crops in Southern Africa region. *Integrated Crop Management*, Vol. 14, FAO, Rome, Italy.

Kossoy, A., Peszko, G., Oppermann, K., Prytz, N., Klein, N., Blok, K., Lam, L.,Wong, L., Borkent, B., 2015. State and Trends of Carbon Pricing 2015 (September). The World Bank, Washington, DC, USA.

Mitchell, D., 2011. Biofuels in Africa: opportunities, prospects, and challenges. The World Bank, Washington, DC, USA.

Msangi, S., Evans, M., 2013. Biofuels and developing economies: is the timing right? *Agricultural Economics*, 44(4–5), 501–510.

Negash, M., Swinnen, J.F.M., 2013. Biofuels and food security: Micro-evidence from Ethiopia. *Energy Policy*, 61, 963–976.

NSO (National Statistical Office), 2012. Integrated Household Survey 2010/11. Lilongwe, Malawi.

NSO (National Statistical Office), 2014. Annual Economic Survey Report 2010–2011. Zomba, Malawi.

Pauw, K., Thurlow, J., Bachu, M., van Seventer, D., 2011. The economic costs of extreme weather events: a hydrometeorological CGE analysis for Malawi. *Environment and Development Economics*, 16, 177–198.

Pauw, K., Schuenemann, F., Thurlow, J., 2015. A 2010 Social Accounting Matrix for Malawi. Unpublished mimeograph, International Food Policy Research Institute, Washington DC, USA.

Quintero, J.A., Cardona, C.A., Felix, E., Moncada, J., Sánchez, O.J., Gutiérrez, L.F., 2012. Techno-economic analysis of bioethanol production in Africa: Tanzania case. *Energy*, 48(1), 442–454.

Rosegrant, M.W., Zhu, T., Msangi, S., Sulser, T., 2008. Global scenarios for biofuels: Impacts and implications. *Review of Agricultural Economics*, 30(3), 495–505.

Rosegrant, M.W. and the IMPACT Development Team, 2012. International Model for Policy Analysis of Agricultural Commodities and Trade (IMPACT): Model Description. International Food Policy Research Institute (IFPRI), Washington, DC.

Searchinger, T., Heimlich, R., Houghton, R.A., Dong, F., Elobeid, A., Fabiosa, J., Tokgoz, S., Hayes, D., Yu, T-H., 2008. Use of U.S. Croplands for Biofuels Increases Greenhouse Gases Through Emissions from Land-Use Change. *Science*, 319(5867), 1238–1240.

SMEC, 2015. National Irrigation Master Plan and Investment Framework – Main Report for Republic of Malawi, Ministry of Water Development and Irrigation, Department of Irrigation. Lilongwe, Malawi.

Thurlow, J., Branca, G., Felix, E., Maltsoglou, I., Rincón, L. E., 2015. Producing Biofuels in Low-Income Countries: An Integrated Environmental and Economic Assessment for Tanzania. *Environmental and Resource Economics*, 64(2), 1–19.

Timilsina, G. R., Beghin, J. C., van der Mensbrugghe, D., Mevel, S., 2012. The impacts of biofuels targets on land-use change and food supply: A global CGE assessment. *Agricultural Economics*, 43, 315–332.

Watson, H. K., 2011. Potential to expand sustainable bioenergy from sugarcane in southern Africa. *Energy Policy*, 39(10), 5746–5750.

Zilberman, D., Hochman, G., Rajagopal, D., Sexton, S., Timilsina, G., 2013. The impact of biofuels on commodity food prices: Assessment of findings. *American Journal of Agricultural Economics*, 95(2), 275–281.

Appendix

Table 3.A1: Model indices, variables and parameters

Indices			
c	Commodities and activities	h	Representative households
f	Factors (land, labor and capital)	t	Time periods
Exogenous parameters (Greek characters)			
α^p	Production function shift parameter	θ^v	Value-added share of gross output
α^q	Import function shift parameter	π	Foreign savings growth rate
α^t	Export function shift parameter	ρ^p	Production function substitution elasticity
β	Household marginal budget share	ρ^q	Import function substitution elasticity
γ	Non-monetary consumption quantity	ρ^t	Export function substitution elasticity
δ^p	Production function share parameter	σ	Rate of technical change
δ^q	Import function share parameter	τ	Foreign consumption growth rate
δ^t	Export function share parameter	υ	Capital depreciation rate
ε	Land and labor supply growth rate	φ	Population growth rate
θ^i	Intermediate share of gross output	ω	Factor income distribution shares
Exogenous parameters (Latin characters)			
ca	Intermediate input coefficients	*pwm*	World import price
cab	Current account balance	*qfs*	Total factor supply
cd	Domestic transaction cost coefficients	*qgov*	Base government consumption quantity
ce	Export transaction cost coefficients	*qinv*	Base investment demand quantity
ci	Capital price index weights	*rf*	Factor foreign remittance rate
cm	Import transaction cost coefficients	*sh*	Marginal propensity to save
cpi	Consumer price index	*tf*	Factor direct tax rate
cw	Consumer price index weights	*th*	Personal direct tax rate
ga	Government consumption adjustment factor	*tm*	Import tariff rate
gh	Per capita transfer from government	*tq*	Sales tax rate
pop	Household population	*wh*	Net transfer from rest of world

pwe	World export price		
Endogenous variables			
AR	Average capital rental rate	*QG*	Government consumption quantity
FS	Fiscal surplus (deficit)	*QH*	Household consumption quantity
IA	Investment demand adjustment factor	*QI*	Investment demand quantity
PA	Activity output price	*QK*	New capital stock quantity
PD	Domestic supply price with margin	*QM*	Import quantity
PE	Export price	*QN*	Aggregate intermediate input quantity
PM	Import price	*QQ*	Composite supply quantity
PN	Aggregate intermediate input price	*QT*	Transaction cost demand quantity
PQ	Composite supply price	*QV*	Composite value-added quantity
PS	Domestic supply price without margin	*WD*	Sector distortion in factor return
PV	Composite value-added price	*WF*	Economy-wide factor return
QA	Activity output quantity	*YF*	Total factor income
QD	Domestic supply quantity	*YG*	Total government revenues
QE	Export quantity	*YH*	Total household income
QF	Factor demand quantity	*X*	Exchange rate

Table 3.A2: Model equations

Prices	
$PM_{ct} = pwm_c \cdot (1+tm_c) \cdot X + \sum_{c'} PQ_{c't} \cdot cm_{c'c}$	1
$PE_{ct} = pwe_c \cdot X_t - \sum_{c'} PQ_{c't} \cdot ce_{c'c}$	2
$PD_{ct} = PS_{ct} + \sum_{c'} PQ_{c't} \cdot cd_{c'c}$	3
$PQ_{ct} \cdot (1-tq_c) \cdot QQ_{ct} = PD_{ct} \cdot QD_{ct} + PM_{ct} \cdot QM_{ct}$	4
$PX_{ct} \cdot QX_{ct} = PS_{ct} \cdot QD_{ct} + PE_{ct} \cdot QE_{ct}$	5
$PN_{ct} = \sum_{c'} PQ_{c't} \cdot ca_{c'c}$	6
$PA_{ct} \cdot QA_{ct} = PV_{ct} \cdot QV_{ct} + PN_{ct} \cdot QN_{ct}$	7
$cpi = \sum_c cw_c \cdot PQ_{ct}$	8

Production and trade	
$QV_{ct} = \alpha^{p}_{ct} \cdot \sum_{f} \left(\delta^{p}_{fc} \cdot QF_{fct}^{-\rho^{p}_{c}} \right)^{-1/\rho^{p}_{c}}$	9
$WF_{ft} \cdot WD_{fct} = PV_{ct} \cdot QV_{ct} \cdot \sum_{f'} \left(\delta^{p}_{f'c} \cdot QF_{f'ct}^{-\rho^{p}_{c}} \right)^{-1} \cdot \delta^{p}_{c} \cdot QF_{fct}^{-\rho^{p}_{c}-1}$	10
$QN_{ct} = \theta^{i}_{c} \cdot QA_{ct}$	11
$QV_{ct} = \theta^{v}_{c} \cdot QA_{ct}$	12
$QA_{ct} = \alpha^{t}_{c} \cdot \left(\delta^{t}_{c} \cdot QE_{ct}^{\rho^{t}_{c}} + \left(1-\delta^{t}_{c}\right) \cdot QD_{ct}^{\rho^{t}_{c}} \right)^{1/\rho^{t}_{c}}$	13
$\frac{QE_{ct}}{QD_{ct}} = \left(\frac{PE_{ct}}{PS_{ct}} \cdot \frac{\left(1-\delta^{t}_{c}\right)}{\delta^{t}_{c}} \right)^{1/\left(\rho^{t}_{c}-1\right)}$	14
$QQ_{ct} = \alpha^{q}_{c} \cdot \left(\delta^{q}_{c} \cdot QM_{ct}^{-\rho^{q}_{c}} + \left(1-\delta^{q}_{c}\right) \cdot QD_{ct}^{-\rho^{q}_{c}} \right)^{-1/\rho^{t}_{c}}$	16
$\frac{QM_{ct}}{QD_{ct}} = \left(\frac{PD_{ct}}{PM_{ct}} \cdot \frac{\left(1-\delta^{q}_{c}\right)}{\delta^{q}_{c}} \right)^{1/\left(1+\rho^{t}_{c}\right)}$	17
$QT_{ct} = \sum_{c'} \left(cd_{cc'} \cdot QD_{c't} + cm_{cc'} \cdot QM_{c't} + ce_{cc'} \cdot QE_{c't} \right)$	18
Incomes and expenditures	
$YF_{ft} = \sum_{c} WF_{ft} \cdot WD_{fct} \cdot QF_{fct}$	19
$YH_{ht} = \sum_{f} \omega_{hf} \cdot \left(1-tf_{f}\right) \cdot \left(1-rf_{f}\right) \cdot YF_{ft} + gh_{h} \cdot pop_{ht} \cdot cpi + wh_{h} \cdot X$	20
$PQ_{ct} \cdot QH_{cht} = PQ_{ct} \cdot \gamma_{ch} + \beta_{ch} \cdot \left(\left(1-sh_{h}\right) \cdot \left(1-th_{h}\right) \cdot YH_{ht} - \sum_{c'} PQ_{ct'} \cdot \gamma_{c'h} \right)$	21
$QI_{ct} = IA_{t} \cdot qinv_{c}$	22
$QG_{ct} = ga_{t} \cdot qgov_{c}$	23
$YG_{t} = \sum_{h} th_{h} \cdot YH_{ht} + \sum_{f} f_{f} \cdot YF_{ft} + t\sum_{c} \left(tm_{c} \cdot pwm_{c} \cdot QM_{ct} \cdot X + tq_{c} \cdot PQ_{ct} \cdot QQ_{ct} \right)$	24

Equilibrium conditions	
$qfs_{ft} = \sum_{c} QF_{fct}$	25
$QQ_{ct} = \sum_{c'} ca_{cc'} \cdot QN_{c't} + \sum_{h} QH_{cht} + QG_{ct} + QI_{ct} + QT_{ct}$	26
$\sum_{c} pwm_{c} \cdot QM_{ct} + \sum_{f} \left(1 - tf_{f}\right) \cdot rf_{f} \cdot YF_{ft} \cdot X_{t}^{-1} = \sum_{c} pwe_{c} \cdot QE_{ct} + \sum_{h} wh_{h} + cab_{t}$	27
$YG_{t} = \sum_{c} PQ_{ct} \cdot QG_{ct} + \sum_{h} gh_{h} \cdot pop_{ht} \cdot cpi + FS_{t} s$	28
$\sum_{h} gh_{h} \cdot pop_{ht} \cdot cpi + FS_{t} x \sum_{c} PQ_{ct} \cdot QI_{ct}$	29
Capital accumulation and allocation	
$AR_{ft} = \dfrac{YF_{ft}}{qfs_{ft}}$	30
$QK_{fct} \cdot \left(\sum_{c'} PQ_{c't} \cdot ci_{c'}\right) = \left(\dfrac{QF_{fct}}{qfs_{ft}} \cdot \dfrac{WF_{ft} \cdot WD_{fct}}{AR_{ft}}\right) \cdot \left(\sum_{c'} PQ_{c't} \cdot QI_{c't}\right)$	31
$QF_{fct+1} = QF_{fct} \cdot \left(1 - \upsilon\right) + QK_{fct}$	32
Land and labor supply, technical change, population growth, and other dynamic updates	
$qfs_{ft+1} = qfs_{ft} \cdot \left(1 + \varepsilon_{f}\right)$	33
$\alpha_{ct+1}^{p} = \alpha_{ct}^{p} \cdot \left(1 + \sigma_{c}\right)$	34
$pop_{ht+1} = pop_{ht} \cdot \left(1 + \varphi_{h}\right)$	35
$ga_{t+1} = ga_{t} \cdot \left(1 + \tau\right)$	36
$cab_{t+1} = cab_{t} \cdot \left(1 + \pi\right)$	37

4 Evaluating irrigation investments in Malawi: economy-wide impacts under uncertainty and labor constraints[1]

Abstract: Irrigation expansion is critical to increase crop yields and mitigate effects from climate change in Sub-Saharan Africa, but the low profitability has led to little irrigation investments in the region so far. Using an integrated modeling framework, we simultaneously evaluate the returns to irrigation arising from both economic and biophysical impact channels to understand what determines the profitability of irrigation in Malawi. Our results confirm that the returns to irrigation cannot cover the costs in Malawi. While labor-intensive irrigation expansion leads to unfavorable structural change in the short-run, the profitability hinges on low irrigated yields that fall far from expectations due to insufficient input use and crop management techniques. On the other hand, we find that the non-monetary benefits of irrigation regarding higher food security, lower poverty, and reduced vulnerability to climate change make investments in irrigation worthwhile to improve the livelihoods of smallholders.

4.1 Introduction

There is still very little irrigation use in Sub-Saharan Africa, despite a large potential in terms of existing water resources in the region (You et al., 2011). While farm-level impact evaluations report higher food security and reductions in rural poverty through irrigation use (e.g. Dillon, 2011; Burney and Naylor, 2012), returns to irrigation are often too low to cover the costs of infrastructure investment, labor and maintenance (Inocencio et al., 2007). Yields of irrigated crops stay far behind their potential and labor-intensive irrigation techniques put pressure on already labor-constrained farmers (FAO, 2006; IWMI, 2007; Woodhouse, 2009). As irrigation is one of the most important means to increase crop yields and alleviate negative effects from climate change (Rosenzweig and Parry, 1994; Nelson et al., 2009), our study seeks to understand what determines the profitability of irrigation in Sub-Saharan Africa (SSA) by holistically evaluating the various returns to irrigation arising from both economic and biophysical impact channels.

1 The article in this chapter is published as Schuenemann, F., Thurlow, J., Meyer, S., Robertson, R. and Rodrigues, J., 2018. Evaluating Irrigation Investments in Malawi: Economy-Wide Impacts under Uncertainty and Labor Constraints. *Agricultural Economics*, doi:10.1111/agec.12412, and is used with permission of the authors.

Benefits and costs of irrigation accrue both directly at the farm level and indirectly through multiplier effects on the rest of the economy (Hussain & Hanjra, 2004). Higher productivity, output, and factor requirements of irrigated agriculture affect other sectors through product and factor markets. Other indirect and rather non-monetary effects include reductions in risk from climate variability for both the individual farmer and the economy as whole (Svendsen et al., 2009). The effectiveness of irrigation is as much dependent on economic factor endowments of land and labor as on agro-ecological conditions and crop management techniques especially in terms of input use. When soils are depleted from nutrients as in many African countries, crops cannot use water efficiently, making fertilizer indispensable (Drechsel et al., 2015). So far, analyses of irrigation in SSA that include both direct and indirect effects are scarce. Only one farm-level impact evaluation of irrigating households examines the risk reduction potential of irrigation vis-à-vis extreme weather events (Dillon, 2011). Previous partial or general equilibrium models that include irrigation and capture multiplier effects on the rest of the economy have mainly looked at issues of water markets and pricing or the efficient use of irrigation water (e.g. Hassan and Thurlow, 2010; Calzadilla et al., 2011; see Dudu and Chumi (2008) for a review). Calzadilla et al. (2013) are the first to measure the impact of increased irrigation under climate change in a global general equilibrium setting for SSA, but do not consider constraints of individual countries.

Our study makes several contributions to the literature on irrigation investments in SSA: Firstly, we develop a computable general equilibrium (CGE) model of Malawi extended with irrigated agriculture to assess both direct and indirect economy-wide impacts of irrigation expansion and changes in cropping intensity on monetary returns, food security and poverty. Secondly, we explicitly analyze the role of labor endowments for the effectiveness of irrigation. While the technology adoption literature cites labor shortages as a typical reason for low adoption rates (e.g. Feder et al., 1985), labor has been mostly neglected in studies on irrigation in SSA despite the fact that labor-intensive irrigation has large economy-wide consequences. Thirdly, we link the CGE model to a process-based crop model to capture the impact of agro-ecological conditions and crop management techniques on irrigated yields. In particular, we analyze the importance of the interaction between irrigation and fertilizer application for increasing crop yields. Finally, we conduct a stochastic uncertainty analysis to assess the potential of irrigation for reducing risk and vulnerability from climate change. To the best of our knowledge, we are the first study to simultaneously capture both monetary and non-monetary returns to irrigation from both economic and biophysical linkages, which will allow us to identify the bottlenecks arising in the various impact channels.

We choose Malawi as a case study country where the government recently launched a major irrigation investment plan to increase agricultural land under irrigation from 2.5% to 10% (SMEC, 2015). Such a large investment is likely to increase agricultural productivity substantially and could add to the already achieved productivity gains under Malawi's fertilizer input subsidy program (FISP) (Arndt et al., 2016). Irrigation expansion could also decrease Malawi's vulnerability to extreme weather shocks in the form of droughts and floods that hurt agriculture and exacerbate poverty and food security (Pauw et al. 2011). Conversely, an increase in labor-intensive irrigation might negatively affect the Malawian economy, which suffers from labor shortages especially in the summer season (Wodon & Beegle, 2006). The following section gives an overview on agriculture and irrigation in Malawi. Section 3 describes our methodology for evaluating irrigation investments based on the different impact channels. Section 4 discusses our simulation results, while Section 5 concludes.

4.2 Agriculture and irrigation in Malawi

Malawi is a small landlocked country in Sub-Saharan Africa that is dominated by agriculture with an employment share of 80% and contributing a third to GDP. Despite economic growth averaging around 5% per annum for the past 10 years, Malawi remains one of the poorest countries in the world with a GDP per capita of 471 USD in 2010 and about half of the population still living below the national poverty line (World Bank, 2017, Pauw et al., 2014). This is because its population of around 15 million people in 2010 is one of the fastest growing in Sub-Saharan Africa with a stable growth rate of about 3% per annum, quickly eating up any gain in total GDP. More than 90% of Malawian farmers are smallholders growing mainly food crops for subsistence consumption (NSO, 2012). The staple food maize amounts to almost 50% of food crop production. Fertilizer input subsidies provided by the government have led to some productivity increases for maize, but many households remain food insecure due to small farm sizes and frequent droughts and floods (Pauw et al., 2011). As Malawi is located in the tropics, most of its average annual 1000 mm precipitation is coming down in the rainy season ranging from November to April with almost no rainfall in the dry season from May to October.

Agricultural land amounts to 60% of Malawi's 9.4 million hectare (ha) land area. Crops make up most of Malawi's agricultural sector and account for 60% of agricultural GDP and around 70% of agricultural land. Table 4.1 gives an overview of crop production technologies in Malawi and their respective production and land shares. The subsistence orientation of the sector is evident with food crops showing a share of 82% of crop GDP. Overall, food crops are more labor intensive than export crops. Rice, fruits and vegetables are the most labor intensive crops and bring a relatively low value-added per worker, but a high value-added per ha of land. Export crops on the other hand exhibit a higher value-added per worker and similar value-added per ha compared to food crops, making them an important driver of growth.

Table 4.1: Agriculture in Malawi

	Production		Labor	Value-added or GDP per unit of input			Irrigation share	
	Area (1000mt)	Yield (mt/ha)	Intensity (people/ ha)	Land (per ha)	Labor (per person)	Share of crop GDP	of land	of GDP
Food crops	3,206.6		0.83	510.5	824.4	0.82		
Maize	1,696.0	2.02	0.30	350.5	1,153.5	0.47	0.001	0.002
Sorghum	143.0	0.59	0.18	68.5	382.5	0.01	-	-
Rice	59.0	1.86	1.07	1,030.2	959.0	0.05	0.114	0.145
Root crops	422.0	4.93	0.23	271.4	1,177.4	0.09	-	-
Pulses	705.0	0.71	0.12	143.4	1,188.0	0.08	0.001	0.001
Vegetables	78.9	4.60	2.9	1,216.8	416.0	0.08	1.0	1.0
Fruit	102.7	10.34	1.00	493.1	494.7	0.04	-	-
Export crops	916.8		0.29	348.2	1,222.2	0.18		
Groundnuts	313.0	0.93	0.09	107.8	1,172.7	0.03	0.001	0.001
Oilseeds	331.0	0.92	0.08	116.5	1,515.1	0.03	0.001	0.001
Tobacco	200.7	1.04	0.41	575.8	1,403.8	0.09	-	-
Cotton	47.0	0.62	0.15	132.2	899.4	0.00	-	-
Sugar	25.2	108.00	0.72	809.0	1,119.8	0.02	1.0	1.0
Other crops	85.2	2.82	0.20	218.3	1,111.1	0.01	0.610	0.620

Source: Own estimates using production data from FAOSTAT; irrigated yield estimates from DSSAT; employment data from IHS3 (NSO, 2012); value-added data from the 2010 social accounting matrix (Pauw et al., 2015) and measured in 2010 US Dollar (2010 exchange rate 150 MKW/$); data on shares of irrigated crop land from AQUASTAT.

Table 4.1 also shows that only a small share of crops is irrigated, even though Malawi has large water resources, with lakes and rivers covering around one fifth of the country. The majority of crops are rainfed and grown during the rainy season in summer, while neighboring countries like Zambia successfully grow irrigated crops during the winter season. Malawi not only misses out on a second season, but the low irrigation intensity means that farmers realize yields much lower than potential yields due to erratic rainfall. The share of land under irrigation currently amounts to 2.5% of cropland with about 100 thousand ha irrigated in 2014 (SMEC, 2015). Irrigation is evenly divided between large-scale estates using mainly formal irrigation systems and smallholders with informal systems, irrigating 48,000 ha and 56,000 ha respectively. More than 70% of estate irrigation schemes consist of large center pivot technology and sprinkler systems on sugar estates. Informal smallholder irrigation of food crops is predominantly characterized by gravity irrigation (56%) and treadles pumps (29%) and water is usually applied by canals and watering cans (SMEC, 2015). Regardless of the technology, water for irrigation comes from a limited number of sources. These include diversions along rivers and water stored in dams, while most informal irrigation water comes from dambos, which are waterlogged depressions containing seepage (SMEC, 2015). Pumping water from lakes is in principal possible, but would require additional power exceeding Malawi's current electricity production capacity. Since electricity in Malawi is produced from a small number of hydropower turbines, Malawi's irrigation potential is dependent on the competing uses of water.

4.2.1 Irrigation master plan

The recently launched irrigation master plan (IMP) for Malawi explicitly considers competing water uses and determines an irrigation potential for Malawi of 400 thousand ha irrigated land of which one fourth is already under irrigation. This was calculated through a comprehensive feasibility analysis that first assessed the biophysical irrigation potential for Malawi considering geography, topography (slope), soil suitability and existing land use (SMEC, 2015). A hydrological feasibility study identified areas with adequate water availability throughout the country, defined as the 80% reliable stream flows after subtracting environmental flow rates and human water needs. The hydrological potential was then matched to the physical criteria as well as water requirements of regional cropping patterns to calculate the potential irrigated area of 400 thousand ha. The subsequent technical feasibility assessment identified irrigation schemes that are either fed through dam storage or dambos, since most possible river diversions are already developed and lake pumping is not yet feasible. Dams are only technically feasible in areas with highland storage potential and lowland agriculture. Scheme sizes range from more than 26,000 ha to only 100 ha depending on water availability and topography. A proposed dam in the Makoko region for example will feed 486 ha and requires a storage reservoir of 5,972,664 m^3 (SMEC, 2015).

The plan recommends low cost irrigation technologies using gravity and treadle pumps to convey the water from the sources and watering cans for application to the field, which is very labor intensive. There is no preference for a certain crop, but both food crops for increased food security and high value exports crops for GDP growth are targeted, produced by both smallholders and large-scale estates wherever suitable. Estimations for investment costs amount to 2,146 million US$ for initial infrastructure and 278 million US$ annual recurrent costs for developing the first 116 thousand ha in the next 20 years (SMEC, 2015). Implementing the whole 300 thousand ha of missing irrigation potential could therefore cost more than 5500 million US$ and take another 20 years. These costs appear prohibitively high for the Malawian economy with a GDP around 6 billion US$ in 2010. On the other hand, Malawi received about one billion US$ annual Official Development Assistance in the last 10 years (World Bank, 2017), suggesting that irrigation investment costs could be paid for by donors. In fact, the plan envisages that 72% of costs are borne by development and (foreign) private partners, 9% by the Malawian government and the rest by farmers themselves, although it is unlikely that Malawian smallholders can contribute more than labor and maintenance costs (SMEC, 2015). Moreover, some of the proposed dams could be used for hydropower production and contribute to the financial feasibility of the irrigation schemes, although the additional costs for hydropower equipment and maintenance are not considered in the IMP.

Even though the high investment cost could be paid for by donors, returns to irrigation in Malawi could stay far behind expectations as was the case in several other Sub-Saharan African countries (Inocencio et al., 2007). We therefore make use of the irrigation potential analysis conducted in the IMP and simulate what the development of irrigation infrastructure will mean for growth, poverty and food security using an integrated modeling framework as explained in the next section.

4.3 Modeling approach

4.3.1 Crop models

Irrigation changes the conditions under which crops are grown and affects the dynamic processes surrounding crop growth above and below ground. To capture these agronomic effects we use the Decision Support Software for Agrotechnology Transfer (DSSAT) cropping system model (Jones et al., 2003). The model suite simulates the growth cycle of crops including changes in soil water and nutrients under specific management practices concerning planting and harvesting, irrigation and fertilizer. Daily weather information is integrated with information from a soil module to compute crop yields. Data on daily weather for Malawi from 1951–2009 is provided at 0.5 degree resolution and generic soil data from the Harmonized World Soil Database at a finer resolution of 5 arc-minute grids. To aggregate yield results from grid to national level, the results for each pixel are weighted according to where crops are effectively grown and their geographic performance using weights from the Spatial Production Allocation Model (SPAM) (You et al. 2011).

We apply the DSSAT model to Malawi and simulate historic yield distributions for 48 years for maize, pulses, groundnuts, oilseeds and vegetables under three different management regimes: crops can be grown either in the rainy summer or in the dry winter season. Planting in the summer season is fixed to take place in November and in April for the winter season, specific sowing dates are computed within the model when soil moisture and temperature are optimal following predetermined rules (Müller & Robertson, 2014). Secondly, crops are either rainfed or irrigated with different intensity. Irrigation systems in the crop model are automatic and set up in a way that once soil moisture falls below a critical level, crops get irrigated with either 0.5 or 1 mm of water. Thirdly, to capture the common utilization of fertilizer in Malawi we vary the application rate of fertilizer, applying either no or 100 kg of nitrogen per hectare. The latter corresponds to the amount of fertilizer Malawian farmers receive under the fertilizer input subsidy program (Arndt et al., 2016).

Table 4.2 reports yield means and standard deviations from the 48 years of crop simulation under the different management options. Without fertilizer, irrigation leads to higher mean yields across all crops, although some crops are clearly more drought resistant than others. Especially oilseeds do not show any substantial yield increases. Standard deviations from the mean are greatly reduced the more irrigation water is applied. The effect of fertilizer on pulses, groundnuts and oilseeds is small since these crops are themselves nitrogen fixing. Yet, for both maize and vegetables, fertilizer is the main driver of yields and reinforces the irrigation effect once more. The simple comparison of yields already indicates that management techniques can be crucial for the success of irrigation. Importantly, every seasonal simulation in the crop model is independent from the other. If crops are grown in the summer and winter seasons on the same soils, soil quality could be deteriorating if not enough fertilizer is applied to replace nutrients. Our results might thus be a bit optimistic, but emphasize the importance of input management and agro-ecological conditions. For the simulations of irrigation expansion, we incorporate the crop model results into an economy-wide model of Malawi.

Table 4.2: Estimated means and standard deviations by crop model crop

	Mean yield (ton per hectare)					
	Without fertilizer			With fertilizer		
Irrigation amount (mm)	**0**	**0.5**	**1**	**0**	**0.5**	**1**
Maize	0.85	0.99	1.09	3.33	4.09	4.73
Pulses	1.14	1.32	1.60	1.26	1.42	1.65
Groundnuts	2.66	2.90	3.19	2.68	2.92	3.22
Oilseeds	5.42	5.58	5.75	5.61	5.73	5.85
Vegetables	0.80	0.84	0.87	2.55	2.68	2.76
Standard deviation of yields (ton per hectare)						
	Without fertilizer			**With fertilizer**		
Irrigation amount (mm)	**0**	**0.5**	**1**	**0**	**0.5**	**1**
Maize	0.24	0.17	0.11	1.04	0.70	0.35
Pulses	0.41	0.43	0.41	0.39	0.40	0.38
Groundnuts	0.42	0.38	0.31	0.41	0.36	0.28
Oilseeds	0.57	0.48	0.37	0.55	0.47	0.36
Vegetables	0.11	0.10	0.08	0.31	0.32	0.33

Source: Own estimates from crop model simulations using DSSAT.

4.3.2 Economy-wide model

CGE models are ideal to measure the economy-wide implications and impact channels of large-scale development policies, since they capture both micro-economic impacts as well as subsequent multiplier effects working through market interactions and sectoral linkages. Our comparative-static CGE model for Malawi follows Lofgren et al. (2002) and builds on previous CGE modeling work on Malawi (Schuenemann et al., 2017) by simulating the functioning of the Malawian economy using behavioral and structural equations that govern the decision-making of economic agents and maintain macroeconomic consistency and resource constraints. Households maximize their utility and producers maximize their profits, both groups interact at factor and product markets where prices adjust to ensure equilibrium. Production in each sector is determined by constant elasticity of substitution (CES) functions that allow substitution between factors based on relative factor price changes. Intermediate input use is calculated using fixed input-output coefficients in Leontief functions.

Commodities can be traded with the rest of the world, with domestic, export and import quantities determined by relative prices (including relevant taxes and transaction costs). Imperfect substitution between imports and domestic goods is governed by a CES function, whereas the quantity of goods exported is based on a constant elasticity of transformation function. World prices are fixed under a small country assumption.

The model's parameters are assigned values derived from a 2010 social accounting matrix (SAM) (see Pauw et al., 2015). The model contains 51 representative production sectors, 26 of which are crops subdivided between irrigated and rainfed production. The different production sectors compete for production factors land, labor, and capital as well as intermediate inputs. The model distinguishes between three types of labor based on education levels. Land is disaggregated between rainfed and irrigated agriculture. As Malawi is highly land and labor constrained especially at the height of the rainy season, production factors in our model are initially fully employed and mobile between sectors. Demand and supply of factors is balanced through the adjustment of factor prices (economy-wide wages and rents).

Production factors are owned by households who get paid wages, rents and profits, which in turn are used to pay for consumption, taxes and savings. To measure the distributional effects of irrigation expansion, the model contains 20 representative households disaggregated by location (rural/urban), farm/nonfarm and income quintile. These households consume according to a linear expenditure system (LES) of demand. Income elasticities are estimated based on Malawi's 2010/11 Integrated Household Survey (IHS3) (NSO, 2012). The impact of irrigation expansion on poverty is evaluated with a top-down microsimulation module in which the representative households from the CGE model are mapped to households from the survey (see Arndt et al., 2012). The module applies consumption changes from the CGE model proportionately to survey households and recalculates each household's consumption levels. The poverty status is determined by comparing the calculated consumption levels to the official poverty line.

To ensure macroeconomic consistency the model is governed by "closure rules". The government acts as a third type of economic agent and earns tax revenues based on fixed tax rates as well as fixed domestic and foreign transfers. These revenues finance fixed levels of recurrent spending, leaving the recurrent deficit to adjust to maintain fiscal balance. The savings-investment balance is maintained through the adjustment of households' marginal propensities to save in order to match fixed real investment. Finally, a flexible exchange rate adjusts to balance the external account. The model's numeraire is the domestic price index.

4.3.3 Simulating irrigation expansion

Crop sectors in the CGE model are disaggregated by irrigated and rainfed production according to output and areas for irrigated and rainfed agricultural land. Data on the share of irrigated land in total crop land for each crop comes from AQUASTAT. Irrigated output from these lands is calculated using the means of irrigated yields from the crop model simulations. The technologies of irrigated crops differ from their rainfed counterparts in terms of labor requirements and maintenance costs. Irrigated export crops grown on estates are produced under the already existing irrigation technologies using large scale sprinkler systems. As intended by the IMP, we assume that half of smallholder irrigation is conducted using treadle pumps and the other half by gravity irrigation. A weighted average of these two technologies shows that they are 1.83 times as labor intensive as rainfed agriculture (SMEC, 2015). Other than the irrigation master plan we assume that all irrigation infrastructure investment costs are born by development partners, while maintenance has to be financed by farmers themselves.[2]

To model the benefits of irrigation arising from non-economic impact channels, we link the crop model results top-down to the CGE model. The crop model translates the historic variation in climate into variations in yield for each crop. The yield distributions directly correspond to historic climate distributions, so that mean yields correspond to a season with average climate. The crop production functions are normalized to this historical mean yield to estimate average impacts of irrigation. The yields further reflect management techniques in Malawi in terms of fertilizer application rates. Currently about half of Malawian maize is fertilized with subsidized fertilizer. Therefore current irrigated maize yields at 2.8 metric tons per hectare are only 40% higher than rainfed yields and much lower than the potential 4.8 tons per hectare using recommended fertilizer rates.

To measure irrigation benefits due to lower risks of climate change impacts we randomly draw climate events from the historical yield distribution and impose

2 We do not explicitly model foreign direct investment by outside donors but technically assume that irrigation infrastructure falls from the sky. This is not perfect but avoids an artificial appreciation of the exchange rate through enormous foreign exchange import, which would bias the modeling results especially in our static model.

them on the CGE model. Following Rodrigues et al. (2016) we convert the distribution of climate affected crop yields $\mathbf{D} \in \mathbb{R}^{n_y \times n_{cr}}$, with n_y and n_{cr} representing the number of years and crops under each management system, into a continuous random variable $\boldsymbol{C_y}$. Since yields of different crops and management types are likely correlated under the same climate, we identify a vector of expected values for each column of $\mathbf{C_y}$ and a covariance matrix that define a multivariate distribution of climate events. The random climate realizations are then drawn from this new distribution and translated into yield deviations from the mean. These yield deviations represent the climate realizations within the CGE model and are imposed on the production functions as a change in crop-level total factor productivity (TFP).[3]

4.3.4 Simulations

We simulate four sets of scenarios as listed in Table 4.3 to accommodate the different impact channels of irrigation and compare the results to a baseline without any new irrigation development. The first scenario measures the general economy-wide effects of irrigation. As potential irrigation areas are exclusively on cultivated land (SMEC, 2015), we increase the land available for irrigation and likewise reduce the land available for rainfed agriculture by 300 thousand ha. Irrigated and rainfed crops do not compete with each other for land, but only for labor, capital and intermediate inputs. In the next two scenarios we simulate an increase in cropping intensity by opening up a second season in winter. Theoretically the land equipped for irrigation can be irrigated all year round, so we run a scenario in which we increase the amount of irrigated land by an additional 300 thousand ha. Since we add new land that was not previously cultivated, there is no need to decrease the amount of rainfed land. We treat the winter and summer season as identical and do not account for potential differences in yield. Under optimal irrigation, yields in both seasons should not be too different from each other but we acknowledge this limitation by running another scenario where we add only 100 thousand ha of new land. This scenario might be more realistic since soil quality could worsen if crops are planted all year round.

---

3 Since not all of the crops included in the CGE model are simulated in the crop model, we estimate historic yields of the missing crops dependent on maize yields as explained in Rodrigues et al. (2016) with correlations shown in Table 4.A1 in the appendix.

Table 4.3: List of scenarios

Name	Sets of scenarios											
	Scenarios with labor constraints			Scenarios with unemployment			Scenarios under uncertainty			Scenarios under uncertainty and fertilizer constraints		
	Summer 300	**Winter 300**	**Winter 100**	**Summer 300**	**Winter 300**	**Winter 100**	**Summer 300**	**Winter 300**	**Winter 100**	**Summer 300**	**Winter 300**	**Winter 100**
Season in which crops are irrigated	Summer season	Winter season and summer season	Winter season and summer season	Summer season	Winter season and summer season	Winter season and summer season	Summer season	Winter season and summer season	Winter season and summer season	Summer season	Winter season and summer season	Winter season and summer season
Amount of land under irrigation	300,000 ha	600,000 ha	400,000 ha	300,000 ha	600,000 ha	400,000 ha	300,000 ha	600,000 ha	400,000 ha	300,000 ha	600,000 ha	400,000 ha
Labor constraints	Yes	Yes	Yes	No	No	No	No	No	No	No	No	No
Weather	Historic average	Historic average	Historic average	Historic average	Historic average	Historic average	Random weather realization	Random weather realization	Random weather realization	Random weather realization	Random weather realization	Random weather realization
Fertilizer on irrigated maize	Yes	Yes	Yes	Yes	Yes	Yes	Yes	Yes	Yes	No	No	No

Source: Own compilation.

In the three scenarios above, we assume that labor is fully employed, as Malawi faces serious labor constraints in the rainy season (Alwang and Siegel, 1999). Apart from generally low capital availability in agriculture, recent studies on agricultural innovation adoption emphasize that labor scarcity is one of the most important reasons why productivity enhancing technologies such as irrigation are not taken up in Sub-Sahara Africa (Woodhouse, 2009; Nin-Pratt et al., 2011). In our simulations, we force farmers to take up the new technology as we take away land from rainfed production. In the short run, an expansion of labor-intensive irrigated agriculture relative to the lower labor requirements of rainfed production may evoke high adjustment costs in terms of structural change. In the long run, a temporary shortage of labor is smoothed out, as the expansion of irrigation is unlikely to happen within a single year. The IMP considers an increase in irrigation to 220 thousand ha within 20 years, whereas in our policy simulations irrigation expands to its maximum potential of 404 thousand hectares. The simulations in our comparative-static model therefore reflect policy impacts after a period of 20 to 30 years. Labor shortages are unlikely to be a problem as Malawi's population is forecasted to grow by 3 percent per annum in the near future (Choi et al., 2016). Moreover, Malawi's labor shortages are of a seasonal nature. Smallholder farmers in the rainy season are very labor constrained due to on-farm labor demands and off-farm labor opportunities, whereas there is often unemployment in the dry season (Wodon and Beegle, 2006). Therefore, the assumption of fully employed labor might not be realistic in the long-run and we repeat the three simulations modeling partial unemployment. This is implemented in the labor market equation by fixing the economy-wide wage rate for low-skilled workers and making the supply of low-skilled labor endogenous. We still maintain full employment for higher skilled labor, reflecting that these labor groups are less affected by seasonality, and thus capture potential effects of irrigation expansion on structural change.

The previous scenarios assess irrigation impacts in an average weather year, but sizeable benefits of irrigation arise from the reduction in risk arising from erratic weather. We repeat the three simulations under uncertainty following the two-step approach used by Rodrigues et al. (2016). We first run the model using the crop production functions based on mean yields. Producers in our model are risk-neutral and take the decision how to allocate their production factors and intermediate inputs assuming average weather and average yields. These resource allocations are fixed and we draw a random weather event from the multivariate distribution of climate events. The deviations from the mean yield evoked by the respective weather are then imposed on the production functions for each crop as a TFP shock and we run the model a second time. The model establishes a new equilibrium by reallocating mobile factors and commodities, which can only be done in the non-farm sectors. By repeating the stochastic simulations 300 times we produce the value distribution of different outcome variables. We also rerun the baseline without irrigation for each random weather draw to establish the correct counterfactual.

Finally, we run a set of scenarios to isolate the impact of biophysical linkages and management techniques for the success of irrigation in increasing yields. Table 4.2 shows the importance of fertilizer in reinforcing the yield increasing effect of irrigation, especially for maize as Malawi's staple crop. In order to isolate the value of the interaction effect between fertilizer and irrigation, we re-run the stochastic simulations under uncertainty without the current fertilizer practices for irrigated maize in Malawi. We use the same two-step procedure as in the climate variability scenarios but irrigated technologies are based on the mean maize yields without fertilizer as shown in Table 4.2.

4.4 Simulation results

To measure the impact of irrigation on the Malawian economy, we compare the simulation results to a counterfactual baseline scenario without any irrigation expansion that maintains the status quo. The first column in Table 4.4 shows the initial and baseline structure of the Malawian economy in the year 2010.

Table 4.4: Production and price impacts

	Initial share or value, 2010	Deviation from baseline value (%)					
		Irrigation scenarios with labor constraints			Irrigation scenarios with unemployment		
		Summer 300 thousand ha	Winter 300 thousand ha	Winter 100 thousand ha	Summer 300 thousand ha	Winter 300 thousand ha	Winter 100 thousand ha
Total GDP	100.00	-0.1	0.8	0.2	1.1	3.8	2.0
Agriculture	0.32	-0.2	1.7	0.4	1.1	4.8	2.4
Food crops	0.17	-0.1	2.1	0.7	1.2	5.3	2.6
Export crops	0.03	0.4	7.2	2.7	1.2	8.9	3.8
Other ag	0.13	-0.5	-0.2	-0.4	0.9	3.3	1.7
Non-agriculture	0.68	-0.1	0.4	0.1	1.1	3.2	1.8
Industry	0.16	-0.5	-1.0	-0.7	0.6	1.7	1.0
Food processing	0.06	-0.7	-0.9	-0.7	1.5	4.4	2.5
Services	0.51	0.1	0.8	0.3	1.3	3.7	2.1
Change in price indices							
Real exchange rate	1.00	0.45	0.97	0.64	0.38	0.83	0.55
Real food prices	1.00	-0.41	-1.05	-0.64	-0.39	-1.00	-0.60
Total crop land (1000ha)	4,209	4,209	4,509	4,309	4,209	4,509	4,309
Food crops	3,520	-1.27	3.23	0.25	-0.39	5.64	1.62
Export crops	689	6.46	27.05	27.05	1.98	14.74	6.24

Source: Results from the Malawi CGE model.

4.4.1 Irrigating summer crops

The second column of Table 4.4 shows the impacts of increasing irrigation by 300 thousand hectares in the summer season on GDP, prices, and cropland allocation. As agricultural land under irrigation is expanded, land planted with rainfed crops is redistributed to irrigated crops. The overall composition of cropland changes as irrigated land planted with export crops displaces some of the former rainfed land used for food production, although the new irrigated land is relatively evenly distributed between food and cash crops. As a consequence GDP from export crop production expands relative to the baseline, while food crop production slightly decreases by 0.1%. Most of the negative effect on agricultural GDP stems from non-crop agricultural activities (livestock, forestry and fishery), from which workers migrate due to high labor demand in the irrigated crop production sectors. As irrigation expands, it not only draws in land and labor from other sectors but also increases demand for maintenance, leading to growth in the services sectors. Other non-agricultural sectors contract as workers migrate into services. Overall impacts on GDP are very small but negative for several reasons. Although irrigated crops have a higher GDP per hectare of land, they have a lower GDP per worker than rainfed crops. Therefore, the overall increase in value-added remains small, as rainfed lands are replaced by irrigated lands. Moreover, the high labor requirements of irrigated agriculture lead to a drain of labor from other productive sectors that contribute to GDP. Thus even though irrigation directly increases crop yields, the structural change happening within agriculture and the non-agricultural sectors has effectively a dampening impact on economic growth. While the production of irrigated high value export crops such as sugarcane and tea increases, production of non-irrigated traditional exports crops such as tobacco decreases. Together with lower industrial exports, this leads to lower total exports compared to the baseline and thus a very small depreciation of the exchange rate. The decrease in food prices does not necessarily lead to improved food security as only prices of irrigated cash crops (vegetables and rice) decrease. All other food and non-food prices increase due to lower food and industrial output.

Impacts on labor and household welfare are reported in the second column of Table 4.5. Although irrigated agriculture is more labor-intensive than rainfed agriculture, the labor share of agriculture does not change as irrigated export crops need less labor on average than the rainfed food crops they displace. The increase in labor demand from the irrigated and services sectors leads to rising wages, which do not translate into higher household welfare. Both farm and non-farm households suffer from higher non-food and food prices for staples as well as lower returns to agricultural capital. As irrigation infrastructure is financed by donors, irrigated agriculture is less capital intensive than rainfed production. An exchange from rainfed to irrigated land therefore reduces demand for capital leading to lower rents. Farm households experience additional income reductions due to higher labor requirements of irrigated agriculture relative to rainfed. Farm households either have to hire additional labor or have to work longer hours on their fields, which reduces their time for off-farm employment. These reductions in income translate into increases in poverty that are larger for farm than for non-farm households.

Table 4.5: Labor and household impacts

	Initial value or share, 2010	Deviation from baseline value (%)					
		Irrigation scenarios with labor constraints			Irrigation scenarios with unemployment		
		Summer 300 thousand ha	Winter 300 thousand ha	Winter 100 thousand ha	Summer 300 thousand ha	Winter 300 thousand ha	Winter 100 thousand ha
Agriculture labor share (%)	63.8	0.0	-0.7	-0.3	0.1	-0.5	-0.1
Real wage	3,696	0.7	1.9	1.2	0.4	1.1	0.6
Household welfare	426	-0.3	0.6	0.0	1.1	3.9	2.1
Farm households	330	-0.4	0.6	0.0	1.2	4.3	2.2
Non-farm households	1,018	-0.1	0.5	0.1	1.0	3.2	1.8
Poverty headcount rate	51.0	0.6	-0.9	0.1	-1.4	-5.3	-2.6
Farm households	48.1	0.6	-0.9	0.1	-1.3	-5.2	-2.5
Non-farm households	2.9	0.1	-2.5	-0.1	-3.4	-7.3	-4.3

Source: Results from the Malawi CGE and microsimulation models.

Notes: The initial value of real wage is average per capita real wage in US$. Welfare is measured using real consumption expenditure, the initial value is average per capita US$ expenditure. Poverty headcount rate is the share of the population with per capita expenditures below the national poverty line.

The second column in Table 4.6 reports investment returns in terms of GDP and absorption gains for the summer irrigation scenario. Absorption encompasses total domestic spending for private and public consumption and investment and captures policy impacts on national welfare. It gives a good measure on the economy-wide benefits of policies through both direct output increases and indirect spillover effects on incomes and will later be used to calculate the economy-wide benefit-cost-ratio (EBCR) of irrigation (Arndt et al., 2016). As expected, GDP from rainfed crops decreases as land is taken away from rainfed production, while irrigated crop GDP is almost three times higher than in the baseline. The impact of irrigation expansion on total crop GDP as well as GDP per hectare is negligible. GDP per hectare of irrigated lands decreases compared to the baseline, in which the majority of irrigated land was planted with high value export crops. When expanding irrigation to 404 thousand hectares a higher share of irrigated land is used for food crop production that exhibit a lower GDP per hectare than export crops. Total absorption falls relative to the baseline mainly due to a decrease in private consumption. Simply expanding irrigation in the summer season is not a profitable option for Malawi. The returns to irrigated land are not large enough to offset losses from reduced rainfed production and higher labor costs for farmers as already indicated by other studies (Woodhouse, 2009; Nin-Pratt et al., 2011). In addition, our economy-wide analysis shows that the competition for limited production factors leads to unfavorable structural change. This means that irrigation expansion does not only directly affect farmers, but also the rest of the economy through consumption and production linkages.

Table 4.6: Investment returns

	Initial value, 2010	Irrigation scenarios with labor constraints			Irrigation scenarios with unemployment		
		Summer 300 thousand ha	Winter 300 thousand ha	Winter 100 thousand ha	Summer 300 thousand ha	Winter 300 thousand ha	Winter 100 thousand ha
Total absorption (million $)	8,167	8,151	8,201	8,168	8,234	8,404	8,292
Total GDP (million $)	6,391	6,384	6,442	6,403	6,461	6,631	6,519
Total Crop GDP (1000 $)	1,260	1,260	1,297	1,273	1,275	1,334	1,295
Rainfed lands	1,198	1,077	1,039	1,063	1,091	1,069	1,082
Irrigated lands	62	183	259	210	184	265	213
Total crop GDP/ha ($)	299	299	288	295	303	296	301
Rainfed lands	292	283	273	279	287	281	285
Irrigated lands	597	452	367	416	456	376	422

Source: Results from the Malawi CGE model.

Notes: All values are measured in US$.

4.4.2 Introducing a winter season

The situation changes once there is an increase in the supply of one factor. Column 3 of Table 4.4 reports GDP, price and land allocation effects if irrigation expansion in summer simultaneously involves an increase in cropping intensity by opening up a second season in winter. Since the amount of available land increases by 300 thousand hectares, competition for land is greatly reduced and all crop production sectors exhibit a higher GDP relative to the baseline. In this scenario, competition for limited labor determines the winners and losers of irrigation expansion. Agricultural GDP is again mainly driven by an increase of high value export crop production. Increased production of food crops, especially maize, on irrigated lands evokes a decrease in food prices. Higher demand for maintenance leads to higher growth in the services sector so that non-agricultural GDP is higher than in the baseline. As demand for labor in the agricultural and services sector increases once more, especially the industrial sector suffers. Industrial output further decreases due to migration of workers into agriculture. Lower industrial exports lead to a larger depreciation of the real exchange rate compared to the summer scenario.

The expansion in output and lower consumer prices translate into benefits for all households that are reported in the third column of Table 4.5. The total labor share of agriculture actually decreases as more workers migrate from unprofitable rainfed agriculture into services. Real wages rise more than in the summer scenario due to a larger increase in labor demand. Farm households exhibit higher welfare increases than non-farm households. When growing crops in the winter season, the productivity of irrigated cropland doubles leading to rising factor incomes for land owners. Higher welfare generates lower poverty. Since non-farm households benefit more from lower food prices than food producing farm households, poverty decreases are larger for non-farm households. This is because even though farm households of all quintiles in Malawi are net buyers of food (NSO, 2012), they still produce a large share of their food consumption on their own fields (for which labor costs have increased) and only benefit from lower food prices for the marketed share of their consumption. In addition, richer farm households in Malawi have a lower subsistence share and a larger marketed share of food consumption compared to poorer farm households, so that lower food prices bring a higher benefit to richer farm households than to poorer. The result that poorer farm households do not benefit much from decreasing food prices due to their small marketed consumption share is in line with findings by Aksoy and Isik-Dikmelik (2008) that higher food prices do not necessarily hurt poorer farmers as a lot of farm households are only marginal net buyers of food.

Rising household incomes translate into higher private consumption leading to higher absorption as reported in the third column of Table 4.6. Once the irrigation infrastructure is used in both the summer and winter season, absorption increases by 33 million US$ per year. Considering the 287 million US$ annual recurrent costs for 116 thousand hectares, an irrigation expansion of 300 thousand hectares would imply about 719 million US$ annual costs (ignoring economies of scale) compared to 33 million benefits. Thus even when initial investment costs are not considered, the EBCR for irrigation in Malawi is smaller than unity. The increases in absorption and crop GDP in this scenario are still higher than in the third and more realistic scenario, where winter irrigation increases by only 100 thousand hectares. The fourth column in Table 4.5 shows that absorption increases by only 1 million US$. Overall this scenario tracks the impacts of the two previous scenarios in showing that, assuming average weather, returns to irrigation in Malawi are small. If labor is limited, the high labor requirements of irrigation have negative impacts on the rest of economy.

4.4.3 Increasing irrigation with unemployment

As previously mentioned, labor shortages are unlikely to be a problem for irrigation expansion in the long-run. We therefore repeat the simulations from above assuming that there is indeed unemployment in Malawi and compare each scenario to the corresponding scenario with labor constraints. In these scenarios there is much less competition for labor but still for land and other factors and inputs. GDP and price impacts are reported in the final three columns of Table 4.4. Once the labor constraint is removed there is an increase of GDP in all sectors. Workers do not migrate to the crop and service sectors as Malawi's factor endowments effectively increase. There is a larger allocation of total cropland to food crops compared to the full employment scenarios due to increases in food demand following increases in income. The latter is also the reason why the decrease in food prices is lower than in the full employment scenarios. In the Winter300 scenario food crop production increases by more than 5% relative to the baseline. As oilseeds are the export crops with the lowest value-added per hectare of land, the increase in food crop land comes almost completely from displacement of export oilseeds, which also leads to a depreciation of the exchange rate.

Households benefit from higher income from land and labor as shown in Table 4.5. As the amount of lower-skilled labor increases with demand, there is only an increase in wages for higher-skilled workers. The more irrigated land comes under production the higher the benefits. In the Winter300 scenario household welfare increases by almost 4% and poverty is reduced by more than

5% relative to the baseline. Again, farm households exhibit higher welfare as the owners of irrigated land, while non-farm households benefit more from lower food prices. The results show that irrigation increases food security both in terms of availability and in terms of access.

In the Winter300 scenario, total absorption increases by 390 million US$ per year as shown in the sixth column of Table 4.6. Compared with the annual recurrent costs of 719 million US$, this amounts to an EBCR of 0.33. To be complete, we also calculate the net present value (NPV) including the 5500 million US$ initial investment costs and using discount factors based on the geometric mean of Malawi's official lending interest rate in the previous 10 years. The latter amounts to almost 20 percent so that even considering a period of 50 years or more, irrigation investments cannot generate a positive NPV. While the assumption that lower-skilled labor simply increases with demand is disputable, our results also show that the profitability of irrigation hinges only partly on the availability of labor. The largest problem is that increases in yields as estimated by the crop model fall far from expectations. Growth in crop output and GDP appear very small compared to the investment costs calculated by the IMP.

4.4.4 Accounting for climate uncertainty

To measure the impact of irrigation under variable weather conditions, we simulate 300 different weather realizations for the three scenarios and the baseline. Irrigation expansion cannot affect the probability of climate events, but it can minimize the risks of climate change impacts by reducing the vulnerability vis-a-vis climate change both at the micro and economy-wide level[4]. Since irrigated crops are more resilient in years with lower than average rain, the impact of such weather conditions on the economy should be less damaging in the irrigated scenarios compared to the baseline. Using the distributions of GDP from the 300 simulations, we compute box-and-whisker plots of the GDP gain in the three scenarios relative to the baseline without irrigation. The latter allows us to compare how an increase in the area under irrigation affects the GDP gain under variable climate. A reduction in risk of climate impacts due to irrigation should not only become apparent in larger means and medians, but mostly through a larger box in the boxplot, as this shows where the middle 50% of the values are located. Figure 4.1 shows the three boxplots of the GDP gain in the three

4 Risks from climate change in our study are very narrowly defined as random deviations from the average weather in Malawi, which are all equally likely, i.e. have the same probability of occurrence.

scenarios, where the "x" denotes the mean value and whiskers represent the lower and upper quartiles of the distribution, with the end of whiskers denoting the minimum and maximum values of the distribution. As expected the median and mean increase with more land under irrigation and are highest in the Winter300 scenario. Similarly, the size of the boxes is larger the more crop area is under irrigation. As irrigated yields are more stable, it is more likely that values fall between the 1st and 3rd quartile of the distribution if irrigation also takes place in the winter season. The boxplots also show that our simulations exhibit more outliers in the positive range, as the positive whiskers are much longer compared to the negative whiskers. All boxes are relatively symmetrical with slight positive skewness, implying a reduction in risk of values falling below the first quartile the more land is under irrigation. Even so, under the Summer300 scenario there is still a slight risk of having a smaller GDP gain with irrigation than without irrigation as a large part of the lowest quartile falls in the negative range. As more land comes under irrigation, this risk diminishes and the Winter300 scenario exhibits exclusively positive values. Overall, the risk reducing effect of irrigation expansion remains relatively small as there is still a relatively large spread considering the length of the whiskers. This is not very surprising as crop model results reveal that irrigated yields in Malawi are on average only 21% higher compared to rainfed crops. Considering that in the Winter300 scenario 16% of total cropland is irrigated, the average increase in yields amounts to only 3.3%. An increase in irrigated crop yields is thus essential to increase the climate risk reducing effect of irrigation in Malawi.

Figure 4.1: Box-whisker-plots showing the GDP gain distribution relative to a non-irrigation baseline

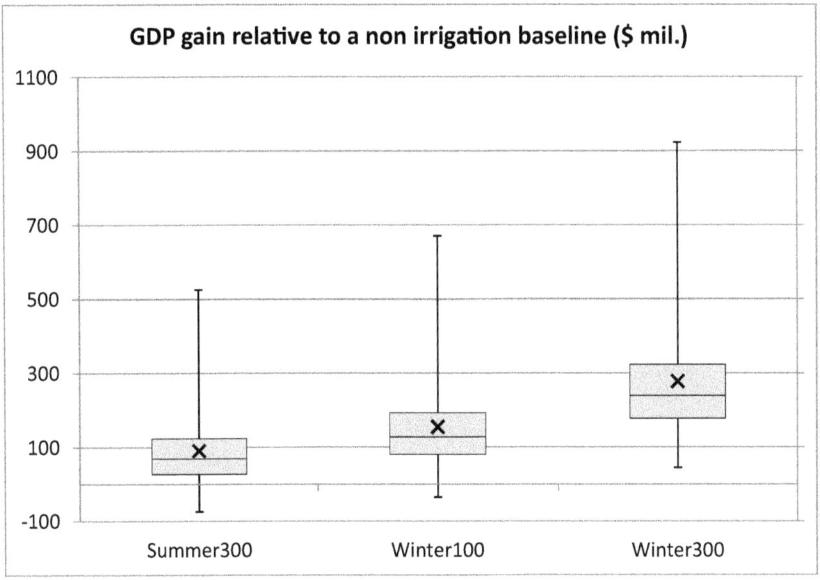

Source: Results from the Malawi CGE model

Notes: Whiskers represent the lower and upper quartiles of the distribution, respectively. End of whiskers denote the minimum and maximum values of the distribution. "x" denotes the mean value. All values are measured in million 2010 US$.

4.4.5 Irrigation fertilizer interaction

Higher yields depend crucially on improved management techniques in terms of fertilizer and water application. Profitable irrigation expansion in Malawi will thus require a continuation and an increase of the fertilizer subsidy program as well as extension services. To better understand the importance of fertilizer for irrigation, we run another set of scenarios where we remove the fertilizer applied to irrigated maize in Malawi. We than compare the GDP gain from our previous scenarios (with fertilizer and irrigation) to the GDP gain without fertilizer (but with irrigation), which should isolate the fertilizer-irrigation interaction effect. Figure 4.2 again shows boxplots that illustrate the distributions of the GDP gain with fertilizer relative to "a with irrigation but without fertilizer scenario" for the three scenarios. The three boxplots show that total GDP gains from irrigation without fertilizer are extremely small and that the likelihood of higher than average outcomes under variable weather is low: in all three scenarios, about

75% of the results fall below the mean. The median GDP gain amounts to only 9 million US$ in the Summer300 scenario and grows to 30 million US$ in the Winter300 scenario, which is about 13% of the median total GDP gain in both scenarios, showing that the majority of the gains from irrigation comes from the interaction between water and fertilizer. Considering that only about half of maize is fertilized in Malawi, fertilizer application and biophysical linkages make thus an essential contribution to increasing crop yields under irrigation.

Figure 4.2: Box-whisker-plots showing the GDP gain distributions relative to a with irrigation but without fertilizer scenario

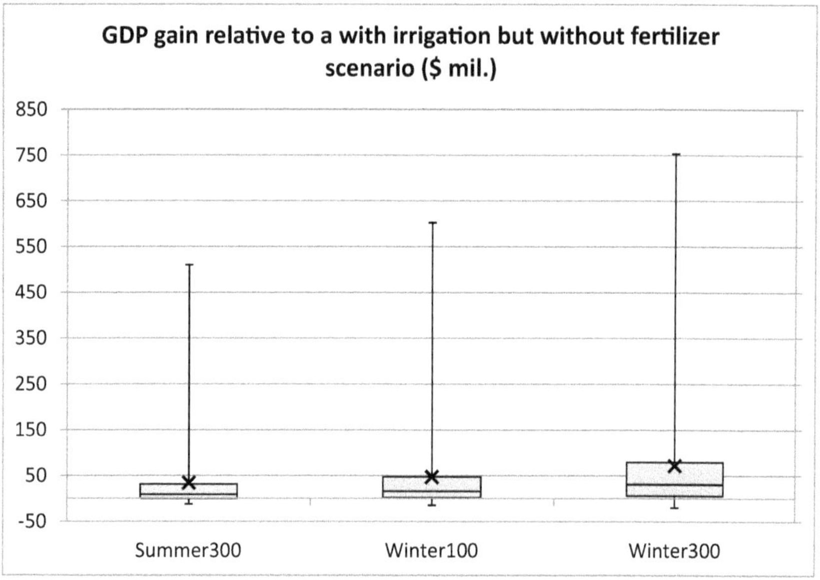

Source: Results from the Malawi CGE model

Notes: Whiskers represent the lower and upper quartiles of the distribution, respectively. End of whiskers denote the minimum and maximum values of the distribution. "x" denotes the mean value. All values are measured in million 2010 US$.

Eventually, our analysis shows that monetary returns to irrigation in Malawi are very small and unlikely to cover the high investment and maintenance cost envisaged by the IMP. As Malawi will continue to suffer from dry spells and climate variability due to climate change that threatens already low crop yields, monetary profitability will not be the main unit to measure returns to irrigation. Much more important benefits arise from increasing food security and decreasing poverty as well as reducing risk and vulnerability to climate change. The question for Malawi is thus not so much whether to invest in irrigation infrastructure or not, but rather how to keep investment costs low, for example through cheaper small-scale irrigation technologies or through multi-purpose dams that can be used for hydropower or to increase tourism.

4.5 Conclusion

Irrigation is an important means for increasing crop yields to reduce food insecurity and vulnerability to climate change in SSA, but the low profitability has led to little investments in irrigation in the region so far. Potential gains of irrigation arise from various impact channels including direct yield increases, indirect multiplier effects and reductions in risk from climate variability. Our study closes an important research gap by measuring both direct and indirect impacts of irrigation investments to understand what determines the profitability of irrigation in SSA. We develop an integrated modeling framework to assess returns to irrigation in Malawi from both economic and biophysical impact channels. Through combining a CGE and a crop model we evaluate the impact of labor endowments, agro-ecological conditions and crop management techniques for the effectiveness of irrigation. Using stochastic simulations we analyze the potential of irrigation to reduce risks of climate change impacts.

Our results show that the returns to irrigation cannot cover the costs in Malawi as calculated by the irrigation master plan. Labor-intensive irrigation expansion leads to unfavorable structural change that has negative impacts on GDP in the short run. Even so, the profitability of irrigation hinges only partly on labor as the decisive factors are irrigated yields that fall far from expectations due to insufficient input use and crop management techniques. Conversely, our simulations also show that irrigation will increase the production of food crops and the welfare of households, leading to lower poverty and higher food security. The climate risk reducing effect of irrigation in Malawi remains small, but positive.

Malawi will continue to be affected by droughts and climate variability that threaten already low crop yields, so that monetary profitability will not be the main unit to measure returns to irrigation. The question for Malawi is therefore not so much whether to invest in irrigation infrastructure or not, but rather how to implement irrigation expansion successfully. Our analysis has shown that biophysical linkages in terms of water-fertilizer interactions play an important role for raising crop yields so that successful irrigation expansion needs to go in hand with improved crop management.

References

Aksoy, M.A. and Isik-Dikmelik, A., 2008. Are low food prices pro-poor? Net food buyers and sellers in low income countries. World Bank, Policy Research Working Paper 4642.

Alwang, J., Siegel, P.B., 1999. Labor Shortages on Small Landholdings in Malawi: Implications for Policy Reforms. *World Development*, 27(8), 1461–1475.

Arndt, C., Pauw, K., Thurlow, J., 2012. Biofuels and economic development: A computable general equilibrium analysis for Tanzania. *Energy Economics*, 34(6), 1922–1930.

Arndt, C., Pauw, K., Thurlow, J., 2016. The Economy-wide Impacts and Risks of Malawi's Farm Input Subsidy Program. *American Journal of Agricultural Economics*, 98(3), 1–19.

Burney, J.A., Naylor, R.L., 2012. Smallholder irrigation as a poverty alleviation tool in Sub-Saharan Africa. *World Development*, 40(1), 110–123.

Calzadilla, A., Rehdanz, K., Tol, R.S.J., 2011.Water scarcity and the impact of improved irrigation management: A computable general equilibrium analysis. *Agricultural Economics*, 42(3), 305–323.

Calzadilla, A., Zhu, T., Rehdanz, K., Tol, R.S.J., Ringler, C., 2013. Economywide impacts of climate change on agriculture in Sub-Saharan Africa. *Ecological Economics*, 93, 150–165.

Choi, N., Lall, S.V., White, R., Webster, M.J., Wang, H.G., 2016. Malawi urbanization review: leveraging urbanization for national growth and development. World Bank Group, Washington, DC.

Dillon, A., 2011. The Effect of Irrigation on Poverty Reduction, Asset Accumulation, and Informal Insurance: Evidence from Northern Mali. *World Development*, 39(12), 2165–2175.

Dixon, P.B., Lee, B., Muehlenbeck, T., Rimmer, M.T., Rose, A. and Vrikios, G., 2010. Effects on the U.S. of an H1N1 Epidemic: Analysis with a Quarterly CGE Model, Journal of Homeland Security and Emergency Management, 7 (1), 1–17.

Drechsel, P., Heffer, P., Magen, H., Mikkelsen, R., Wichelns, D. (Eds.) 2015. Managing Water and Fertilizer for Sustainable Agricultural Intensification. International Fertilizer Industry Association (IFA), International Water Management Institute (IWMI), International Plant Nutrition Institute (IPNI), and International Potash Institute (IPI). First edition, Paris, France.

Dudu, H., Chumi, S., 2008. Economics of Irrigation Water Management: A Literature Survey with Focus on Partial and General Equilibrium Models. World Bank Policy Research Working Paper 4556. Washington, DC: World Bank.

Elliott, J., Deryng, D., Müller, C., Frieler, K., Konzmann, M., Gerten, D., et al., 2014. Constraints and potentials of future irrigation water availability on agricultural production under climate change. In: Proceedings of the National Academy of Sciences (PNAS), 111 (9), 3239–3244.

FAO, 2006. Demand for products of irrigated agriculture in sub-Saharan Africa. FAO Water Reports, 31, Rome, Italy.

FAO, 2013. Atlas of Malawi Land Cover and Land Cover Change 1990–2010. Rome, Italy.

FAO. 2015. FAOSTAT Online Database. Rome, Italy: Food and Agriculture Organization. Available at www.faostat3.org – last accessed 30 April 2015.

Feder, G., Just, R., Zilberman, D., 1985. Adoption of Agricultural Innovations in Developing Countries: A Survey. Econ. Devel. Cult. Change, 33 (2), 255–298.

Hassan, R., Thurlow, J., 2011. Macro-micro feedback links of water management in South Africa: CGE analyses of selected policy regimes. *Agricultural Economics*, 42(2), 235–247.

Hussain, I., Hanjra, M.A., 2004. Irrigation and Poverty Alleviation: Review of the Empirical Evidence. *Irrigation and Drainage*, 53, 1–15.

Inocencio, A.; Kikuchi, M.; Tonosaki, M.; Maruyama, A.; Merrey, D.; Sally, H.; de Jong, I. 2007. Costs and performance of irrigation projects: A comparison of sub-Saharan Africa and other developing regions. IWMI Research Report 109, International Water Management Institute, Colombo, Sri Lanka.

IWMI, 2007. Water for Food, Water for Life: A Comprehensive Assessment of Water Management in Agriculture. London: Earthscan, and Colombo: International Water Management Institute.

Jones J.W., G. Hoogenboom, C.H. Porter, K.J. Boote, W.D. Batchelor, L.A. Hunt, P.W. Wilkens, U. Singh, A.J. Gijsman, and J.T. Ritchie. 2003. The DSSAT cropping system model. *European Journal of Agronomy*, 18(3–4), 235–265.

Lofgren H., R.L. Harris and S. Robinson. 2002. A Standard Computable General Equilibrium (CGE) Model in GAMS. Washington D.C.: IFPRI.

Müller, C. and R.D. Robertson. 2014. Projecting future crop productivity for global economic modeling. *Agricultural Economics*, 45(1), 37–50.

Nelson, G. C., Rosegrant, M. W., Koo, J., Robertson, Richard D., Sulser, T., Zhu, T., Ringler, C., Msangi, S., et al., 2009. Climate change: Impact on agriculture and costs of adaptation. Food Policy Report, International Food Policy Research Institute Washington, D.C.

Nin-Pratt, A., Johnson, M., Magalhaes, E., You, L., Diao, X., Chamberlin, J., 2011. Yield Gaps and Potential Agricultural Growth in West and Central Africa. IFPRI research monograph, Washington, D.C.: International Food Policy Research Institute (IFPRI).

NSO (National Statistical Office). 2012. Integrated Household Survey 2010/2011. Lilongwe, Malawi.

Pauw, K., Thurlow, J., Bachu, M., Van Seventer, D.E., 2011. The economic costs of extreme weather events: a hydrometeorological CGE analysis for Malawi. *Environment and Development Economics*, 16(2), 177–198.

Pauw, K., Schuenemann, F., Thurlow, J., 2015. A 2010 Social Accounting Matrix for Malawi. Unpublished mimeograph, International Food Policy Research Institute, Washington DC, USA.

Pauw, K., Beck, U., Mussa, R., 2014. Did rapid smallholder-led agricultural growth fail to reduce rural poverty? Making sense of Malawi's poverty puzzle. WIDER Working Paper 2014/123.

Rodrigues, J., Thurlow, J., Landman, W., Ringler, C., Robertson, R., Tingju, Z., 2016. The economic value of seasonal forecasts stochastic economywide analysis for East Africa. IFPRI Discussion Paper 1546. Washington, D.C.: International Food Policy Research Institute (IFPRI).

Rosenzweig, C., Parry, M.L., 1994. Potential impact of climate change onworld food supply. *Nature*, 367, 133–138.

SMEC, 2015. National Irrigation Master Plan and Investment Framework for the Republic of Malawi, Ministry of Agriculture, Irrigation and Water Development, Department of Irrigation.

Svendsen, M., Ewing, M., Msangi, S., 2009. Measuring Irrigation Performance in Africa. IFPRI Discussion Paper 00894.

UN, 2015. General Assembly resolution A/70/L.1, Transforming our world: the 2030 Agenda for Sustainable Development.

Wodon, Q., and K. Beegle. 2006. Labor Shortages Despite Underemployment? Seasonality in Time Use in Malawi. In Gender, Time Use and Poverty in Sub-Saharan Africa, Blackden, C.M., and Q. Wodon eds., 97–116. Washington DC: World Bank.

Woodhouse, P., 2009. Technology, Environment and the Productivity Problem in African Agriculture: Comment on the World Development Report 2008. *Journal of Agrarian Change*, 9(2), 263–276.

World Bank. 2017. World Development Indicators Online Database. Washington DC, USA.

You, L., Ringler, C., Wood-Sichra, U., Robertson, R., Wood, S., Zhu, T., Nelson, G., Guo, Z., Sun, Y., 2011. What is the irrigation potential for Africa? A combined biophysical and socioeconomic approach. *Food Policy,* 36(6), 770–782.

Appendix

Table 4.A1: Crop yield correlations with maize yields, 1961–2013

	Coefficient (t-statistic)		R^2	N
Other cereals	0.540	(13.23)*	0.62	265
Root crops	0.253	(4.80)*	0.98	265
Pulses and oilseeds	0.152	(5.02)*	0.94	265
Horticulture	0.081	(2.10)**	0.98	265
Tobacco	0.223	(2.88)*	0.27	265
Cotton	0.210	(3.57)*	0.48	263
Sugarcane	0.598	(8.73)*	0.99	253
Other crops	0.328	(7.86)*	0.79	257

Source: Own calculations using FAOSTAT data (FAO 2015) and based on Rodrigues et al. (2016).

Notes: * and ** indicate significance at the one and five percent level, respectively.

5 Policies for a sustainable biomass energy sector in Malawi: enhancing energy and food security simultaneously[1]

Abstract: Biomass energy still dominates the energy sector in Sub Saharan Africa, in particular as the main cooking energy source in rural and urban areas. The strong linkages to food security and the environment place biomass energy at the heart of sustainable development, a fact that is largely ignored by policy makers in favor of modern energy. At the same time, population and GDP growth are exacerbating already existing supply-demand imbalances in highly populated countries such as Malawi. These trends make it imperative to identify policy interventions that promote sustainable biomass energy while simultaneously considering linkages with other sectors. We use new data on demand and supply for biomass energy in Malawi and develop a model that estimates fuelwood demand based on actual diets and project demand in future years. We simulate how demand side interventions in the form of improved cookstoves affect biomass demand and built a behavioral model to analyze the potential of agroforestry for promoting a sustainable biomass energy sector in Malawi. Our findings show that policy measures aimed at increasing cooking efficiency are not enough to decrease demand for cooking energy due to high population growth. Supply side interventions like agroforestry on the other hand will not only increase sustainable supply, but can also enhance food security and protect the environment. We find that biomass energy can be inherently sustainable and should be an integral part of every energy sector strategy in developing countries as well as of the Sustainable Development Goals.

5.1 Introduction

Sustainable energy is a key focus of the Sustainable Development Goals (SDGs) with SDG 7 seeking to secure access to renewable, "sustainable and modern energy" (UN, 2015). Sub-Saharan Africa's energy sector remains to be dominated by solid biomass in the form of firewood and charcoal, which as renewable but traditional sources of energy are not targeted by the SDGs.[2] SDG 7 rather

1 The article in this chapter is previously published and is used with permission from Schuenemann, F., Msangi, S., and Zeller, M., 2018. Policies for a Sustainable Biomass Energy Sector in Malawi: Enhancing Energy and Food Security Simultaneously. *World Development*, Vol. 103, 14–26, Elsevier, 2018.

2 LPGs and petroleum are also prevalent energy sources in urban areas of Sub-Saharan Africa (IEA, 2014), but are neither targeted by SDG 7 as non-renewable fossil fuels.

aims at universal electrification with electricity produced from renewable sources such as hydro, wind, solar, but biomass and biogas are also promoted as feedstock for generators (UNDP, 2016). Even though SDG 7 seeks a departure from solid biomass energy, including biomass in the SDGs is an important signal to countries in Sub-Saharan Africa. Wood-based energy has long been perceived as backward and harming the environment and studies making the point for renewable and sustainable biomass energy remained largely unheard by policy makers (e.g. World Bank, 2011; Owen et al., 2013).

This is a dangerous oversight considering that almost 75% of people in Sub-Saharan Africa rely on biomass energy for their daily food intake (IEA, 2014), making biomass a decisive factor for SDG 2 "achieving food security and improved nutrition" (UN, 2015) in terms of food utilization. Moreover, sustainable biomass energy is at the heart of SDGs 15 and 13 focusing on sustainable forest management and climate change mitigation. These strong linkages of biomass with other sectors make it imperative to ensure sustainable biomass energy while simultaneously tackling the goal of more modern energy sources. In light of unprecedented pressure on scare resources through population and economic growth, overcoming the tendency to solely focus on a single sector is crucial to find policy measures that minimize trade-offs and increase synergies for all sectors. Our study aims at fostering integrated development policy by seeing the biomass energy sector from a holistic point of view, considering the linkages with the environment, the economy and social well-being, with particular attention to food security.

Malawi makes an important case study for energy sector development in Sub-Saharan Africa in several ways. It is one of the poorest and fastest growing countries in Africa in terms of population with a heavy reliance on biomass energy for food security, since biomass is the predominant cooking fuel of 97% of the population, affecting the utilization dimension of food security (NSO, 2012). There are already divergences between demand and supply of biomass energy in the Southern region of Malawi (Owen et al., 2009). Due to high population growth and total GDP growth at around 5% per year, this gap could be aggravated in the coming decades if there is increased demand for food and cooking energy without any fuel switching to higher value fuels such as LPGs. These trends will also reduce supply indirectly through land clearing as a consequence of growing demand for agricultural land.

In the absence of policy action not only are energy and food security placed at risk but also environmental quality – measured through the impact on forest ecosystem services such as watershed protection and carbon sequestration.

Demand-side measures to increase cooking efficiency have already been initiated but might not be enough considering the projected growth in population. Supply-side measures such as reforestation are inhibited by conflicting land uses for forestry and agriculture. These conflicts could be avoided, if rural households produce their own wood simultaneously with food on their fields. Agroforestry has the potential to increase biomass energy supply, while protecting forests and increasing food security through fertilizer trees (Garrity et al., 2010). Yet, agroforestry policies will mean additional work for already labor-constrained farmers and might not bring the intended benefits if the constraints facing farmers are not taken into account. Our study analyzes how demand and supply for biomass energy in Malawi will evolve in the coming years and which policy measures could ensure a sustainable biomass energy sector. We develop a new methodology with minimum data requirements to estimate demand based on actual diets and cooking habits and project future demand following GDP and population growth. This methodology is expanded to capture effects of efficiency increases in cooking appliances on energy demand. Moreover, we develop a behavioral model for agroforestry adoption considering constraints of rural households in Malawi to analyze the actual potential of agroforestry. Our goal is to show that biomass energy can foster sustainable development and should be an inherent part of energy sector strategies in developing countries.

The following section briefly examines the energy sector in Malawi and the ongoing trends. Section three explains our methodology for estimating and projecting demand for biomass energy as well as results for future demand, section four examines supply side estimations. Section five analyzes the potential of demand side policies in the form of improved cookstoves to decrease demand for biomass energy. Section six explains our model for agroforestry in Malawi and explores how a supply side measure could promote a sustainable biomass energy sector, while section seven concludes.

5.2 Background: energy sector development in Malawi

As in most developing countries in Sub-Saharan Africa, Malawi's energy sector is dominated by biomass. In 2010, 97% of households used biomass energy in the form of firewood, charcoal and crop residues as their main fuel for cooking (NSO, 2012). Biomass constitutes 90% of total energy use, while other energy sources continue to play a minor role (Owen et al., 2009): Electricity use is limited mainly to the sugar industry and to urban areas with only 8% of households being connected to the grid (NSO, 2012). Imported fossil fuels are expensive and rarely used except for the transportation sector. Therefore, biomass will clearly remain the dominating energy source in Malawi in the near future, but is faced with increasing divergences between demand and supply.

Several studies have analyzed Malawi's biomass energy sector and warned about diminishing forest cover for decades (e.g. Orr et al., 1998; Zulu, 2010). These studies underestimated supply by omitting trees outside forests and overestimated demand by overlooking fuel substitution behavior of households from firewood to more inferior energy sources such as crop residues. More recent studies paint a less pessimistic but still alarming picture: Owen et al. (2009) come to the conclusion that sufficient supply of fuelwood is given on a macro level, but find large regional imbalances already in 2008, especially in the overpopulated South where demand exceeds sustainable supply by 10%. A more recent forest valuation study by Hecht and Kasulo (2013) calculates that demand for firewood in the Southern region in 2010 exceeded sustainable yield of forests by a factor of 5, but omitted trees outside forests as well as alternative biomass energy sources. Regardless of the actual magnitude of deviations, the sustainable supply-demand situation is likely to be aggravated in the coming decades by several trends as examined in the following.

5.2.1 Population growth, income growth and urbanization

The household sector represents the largest consumer of biomass energy in Malawi with a share of 92% of total demand, the rest is made up of a few industries such as tobacco processing and brick burning (Owen et al., 2009). Within the household sector, biomass energy is almost exclusively used for cooking, heating and water boiling. Malawi's population is growing rapidly at 3% per annum, urban growth rates are slightly higher with 4% (Choi et al., 2016). At the same time, GDP is growing at around 5% each year, increasing incomes and the demand for food and cooking energy. These developments will not only increase overall demand for biomass energy, but the composition of energy mix is likely to change as people move to urban areas and demand more urban fuels, predominantly charcoal. Higher charcoal consumption will mean an over proportional increase in demand for wood due to conversion inefficiencies. While these trends increase demand for wood directly, they also reduce supply indirectly by increasing demand for land for agriculture. Land clearing for agriculture remains the main reason for deforestation and recent analyses of land cover change in Malawi show that forest cover decreased by the same share by which agricultural land increased in the last 20 years (FAO, 2013).

Figure 5.1 illustrates the effects of population and GDP growth on demand and supply of biomass in a simple model. On the demand side increasing demand for food $\boldsymbol{D}_{food}$ (quadrant I) directly increases demand for cooking energy $\boldsymbol{D}_{cook}$ (quadrant II), which translates into increased demand for biomass $\boldsymbol{D}_{wood}$ depending on the cooking technology used (quadrant III). While traditional cooking appliances such as the three-stone fire need a lot of firewood, improved cook stoves are more efficient and can significantly reduce the demand for biomass energy. In quadrant VI, demand for biomass comes together with sustainable supply $\boldsymbol{S}_{sustainable}$, which is in turn reduced by population and GDP growth through land clearing for agriculture. For illustration, we assume that demand for biomass using a traditional cook stove $\boldsymbol{D}_{wood0}$ and sustainable supply of biomass are in equilibrium in 2010 at point $\boldsymbol{A}_0$, but will be in disequilibrium once population and GDP growth increase the demand for food beyond the level of 2010. Using an improved cooking technology rotates the wood demand curve to $\boldsymbol{D}_{wood1}$, leading to a new equilibrium at point $\boldsymbol{A}_1$ and ensuring equilibrium beyond 2010. However, as soon as population and GDP growth exceed point $\boldsymbol{B}_0$, demand for biomass exceeds sustainable supply even with improved cooking technologies. After this point, only supply side measures such as agroforestry can mitigate the increase in demand by shifting the supply curve outward to $\boldsymbol{S}_{agroforestry}$ and can ensure sustainable fuelwood supply (at least up until point $\boldsymbol{B}_1$). In the following sections we will thus analyze the potential of these two policy options to establish a sustainable biomass energy sector in Malawi.

Figure 5.1: Impact of population and GDP growth on demand and supply of biomass

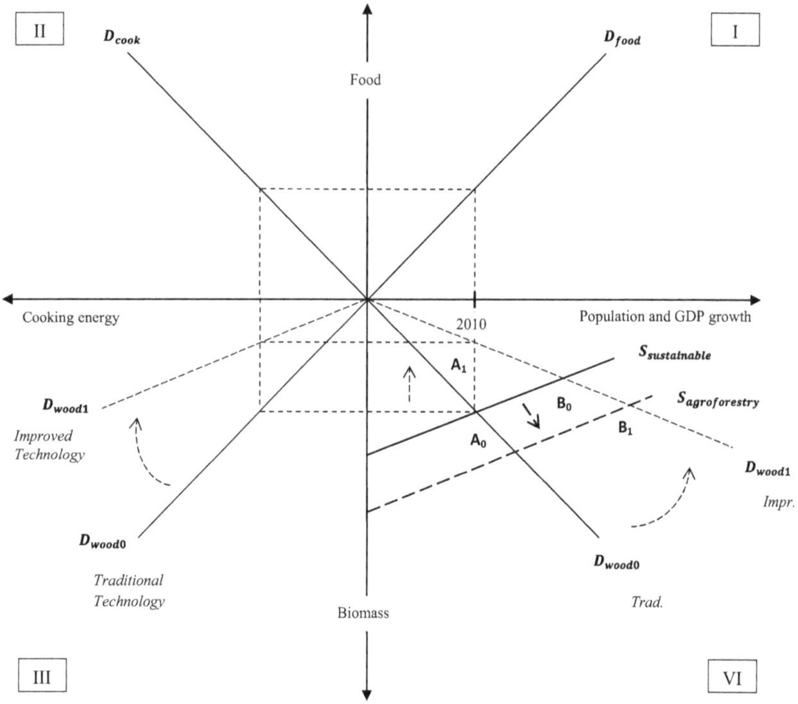

Source: Author's creation

5.3 Estimating biomass energy demand

Finding policy measures for a sustainable biomass energy sector in Malawi requires accurate numbers on future demand given population and GDP growth. While non-household demand is relatively small and can be estimated from existing resources, household energy demand may vary substantially over locations and income groups. We therefore develop a new methodology that indirectly estimates biomass energy household demand through food consumption.

5.3.1 Estimating household demand

Estimations of rural household biomass energy demand in Malawi so far were based on small surveys and then extrapolated for the whole country (Owen et al., 2009) or used secondary data for averages of annual fuelwood consumption (Hecht & Kasulo, 2013). Reliable urban energy consumption surveys have not been conducted in Malawi since 1997 and all recent data are updates from this survey. In the absence of primary information on actual biomass energy consumption values we make use of the direct relationship between demand for cooking energy and food consumption. All meals that the household consumes have to be cooked and this will determine the amount of cooking energy demanded. It is thus possible to directly calculate the demand for cooking energy based on the food consumed. Speaking in economic terms, cooked food and cooking energy are perfect complements with L-shaped indifference curves and Leontief utility functions. In this sense we refrain from taking the approach used in many microeconomic studies which model the amount of firewood collected as dependent on the consumers' preferences for time allocation between productive activities, leisure and firewood collection (e.g. Chen et al., 2006; Murphy et al., 2015). The consumer will always collect so much firewood as it is necessary to cook the food available or cook so much food as is possible with the energy available. Food security and energy security are thus dependent on each other and any increase in the access dimension of food will lead to increased demand for energy, while the latter is essential for the utilization dimension of food security. If more firewood is collected than needed, this is sold at the market and is captured as another household's demand who buys the wood at the market.

To calculate baseline demand for biomass energy, we use data from the Malawi Integrated Household Survey (IHS) 2010/11, a nationally-representative survey providing information on weekly food consumption and on basic cooking energy demand patterns (NSO, 2012). To get a reasonably detailed sample we divide survey households into 30 representative types according to expenditure quintile, region and location (rural/urban). Within these types, a further subdivision is done according to main cooking fuel (firewood, charcoal, electricity, LPGs). We calculate the average diet in kg of each representative household type, divided into 12 broad food groups, 9 of which are cooked directly by the household. Malawi's staple food maize dominates food consumption of all household groups, with a share between 50–90% of total consumed food weight depending on income quintile. Overall, diets of different income quintiles in Malawi are not very different from each other except for in quantity as already found by Tschirley et al. (2015).

To assess how food consumption translates into biomass energy demand for cooking, we make use of Controlled Cooking Tests measuring fuel consumption of local meals cooked by local cooks using different stove types (Bailis et al., 2007). These tests measure the efficiency of stoves with specific fuelwood consumption (SFC), the ratio between the quantity of fuelwood consumed per meal (kg) and the weight of food before cooking (kg), which is ideal to calculate biomass energy needs of diets. Table 5.A1 in the appendix shows the SFC for the 9 food groups and their respective sources for the three stone fire and two improved stoves, the local *Chitetezo Mbaula* and the *rocket* stove. Since most tests are conducted only for firewood and not for charcoal, the charcoal SFC is calculated in wood equivalent by weighing wood SFC with the charcoal energy value and the conversion efficiency of wood into charcoal. The *SFC* values for each *food* group, *fuel* type and *stove* are then multiplied with the consumption quantities $QF_{h,food}$ of the different food groups for each representative household *h*, giving the baseline biomass energy consumption for cooking $QE0_{h,fuel,stove}$ of Malawian households in 2010 as shown in equation (1):

$$QE0_{h,fuel,stove} = \sum_{food} SFC_{food,fuel,stove} * QF_{h,food} \quad (1)$$

Biomass energy is not only used for cooking but for water boiling and heating as well. We rely on findings from Owen et al. (2009) that 76% of household energy in Malawi is used for cooking and the rest for water boiling and heating, and weigh the final value from the cooking module accordingly. Since biomass plays virtually no role for lighting, we exclude lighting from our analysis.

5.3.2 Household demand projections

To analyze which policy measure could establish a sustainable energy sector in Malawi, it is necessary to understand how demand will develop in the light of GDP and population growth. We choose a projection time frame of 20 years, from 2010–2030. Population projections amount to 3% growth per annum nationwide, with higher growth in urban areas (4%) than in rural areas (Choi et al., 2016). We assume Malawi will continue on its growth path of 5% GDP growth per annum for the next 20 years, a very optimistic assumption but in line with growth projections (World Bank, 2016). This will in turn increase incomes (although not by the same rate due to population growth) and thus food and energy consumption.

The indirect estimation of bioenergy demand has several advantages for demand projections since we only need information on consumer preferences for food and not for energy consumption. Our method does not require any estimation of shadow prices for collected biomass energy and is completely independent of market prices of energy, as we explicitly assume that biomass energy demand is not sensitive to own price changes. Although cooking energy demand is also completely inelastic to income changes, we capture the income effect on energy demand implicitly as income growth will change food demand

and cooking energy demand accordingly. The latter effect is measured through a linear expenditure system (LES) derived from a Stone-Geary utility function, which we estimate for each representative household group using consumption expenditure data from the IHS. Under the demand regime dictated by the LES functional form, households consume goods as fixed shares of their supernumerary income, i.e. the income available after subsistence consumption. The equations that describe the LES demand system can be found in the appendix in Table 5.A2. Income elasticities of demand for the different household groups are estimated following King and Byerlee (1978). All LES parameters are thus differentiated across the different household groups so that income changes affect food demand differently according to their specific diets, income elasticities of demand and total expenditure. We use earlier findings from a computable general equilibrium (CGE) model to estimate how the projected population and GDP growth affects income of the 30 representative household groups from 2010–2030 (Schuenemann et al., 2017). A detailed explanation of the Malawi CGE model and the impacts of GDP growth on different household groups can be found in Schuenemann et al. (2017). The yearly income changes are applied to the LES leading to consumption changes. The changes in food consumption are then imposed on the consumption quantities in equation (1). By adding a time dimension we measure the effect on biomass energy consumption in every year.

Treating energy demand as independent from income and prices of other energy sources is a very strong assumption on consumer preferences and implies that households will not switch to other fuels from their baseline fuel if they can afford it. This contradicts the energy ladder hypothesis, which assumes that households prefer modern over traditional fuels (e.g. LPGs over biomass and Electricity over LPGs) and switch once their incomes increase (Hiemstra-van der Horst & Hovorka, 2008). On the other hand, most empirical studies find a weak relationship between fuel source and income in developing countries (van der Kroon, et al., 2013). This is also the case in Malawi, where biomass energy is the main cooking fuel of all income quintiles and fuel "stacking" is widely prevalent in urban areas in Sub Saharan Africa, where different fuels are used for different purposes even if modern fuels are available and affordable (IEA, 2014). We thus do not assume any increase in use of a particular fuel due to fuel switching but only through population and income growth. Since we account for urbanization through higher population growth in urban than in rural areas, there will be an automatic higher growth in demand for those fuels preferred by urban households relative to firewood use in rural areas. Our assumptions are in line with projections of the African Energy Outlook that does not assume any fuel switching in rural areas (IEA, 2014). While the IEA projections consider substantial fuel switching in urban areas from biomass fuel to LPGs, this is not a feasible assumption for Malawi, since LPGs play virtually no role (Owen et al., 2009). There are multiple reasons why LPGs are unlikely to become more important in urban Malawi in the future, including prohibitively high costs of imported LPGs and of LPG cooking appliances, price and market fluctuations but also the big size of the LPG canisters (MERA, 2016). Especially poorer households cannot afford to purchase large quantities and usually buy small amounts of charcoal and firewood on a daily basis (Kambewa et al., 2007). Charcoal remains the dominant fuel for the urban population in Malawi as it is culturally preferred, less expensive, more reliable and available than electricity and LPGs (MERA, 2016). While Malawi's GDP growth is projected at 5% per annum, the high population growth means that per capita income growth remains very small and unlikely to increase the affordability of LPGs and their appliances.

In our calculations, we consider that some households already use an improved cookstove (ICS) and have consequently lower fuel needs. ICS projects have been ongoing in Malawi for several years. Until early 2014, about 165,410 ICS of different fuel types were distributed through several projects, which omits ICS in urban and semi-urban areas that had been adopted prior to any projects (Minofu & Kunert, 2014). We thus assume a best guess conservative estimate of 150,000 adopted ICS in 2010, which corresponds to 5% of Malawian households. This number might be too high considering that the scaling up in some projects happened between 2011 and 2014, but should also reflect non-project ICS in urban areas. In a final step we model how population growth influences energy demand growth. Total fuel use for each year for each representative household group is multiplied with actual population numbers considering population growth and urbanization.

5.3.3 Non household demand

Household biomass energy demand accounts for the majority of fuelwood demand, but several industries are still relying heavily on wood. Tobacco curing as well as brick burning dominate industrial woodfuel demand with a share of 85%. Tobacco is the most important export crop of Malawi and the country produces mostly air-dried Burley tobacco, but about 10% of tobacco is flue-cured. Flue-cured tobacco requires a large amount of wood of around 12 tons of wood per ton of green tobacco (Bunderson & Hayes, 1997). Using output numbers of 173 thousand tons of tobacco for 2010 results in almost 500 thousand tons of wood for tobacco processing.

For the calculation of brick burning fuel demand, we follow the approach used by Hecht and Kasulo (2013). We calculate the average number of new brick houses built annually in the last 6 years, which roughly follow household population growth. The IHS 2010/11 gives information about the number of rooms of each new house and we use house sizes from Ngoma and Sassu (2004) to calculate the amount of bricks necessary to build the houses. Since 0.9 tons of firewood are needed to produce 1000 bricks (Makungwa, 2008), around 500 thousand tons of firewood are required in 2010 for brick burning. Other minor activities consuming biomass energy are cottage industries and the tea sector requiring firewood for tea drying as well as institutions such as schools or hospitals, restaurants and food processors (Owen et al., 2009). The timber industry competes for wood to produce poles and sawnwood. Baseline values for minor activities come from data collected by Owen et al. (2009) for 2008 and are extrapolated to 2010. We assume that all minor industries grow in line with GDP growth. Brick burning and timber production follow population growth as was the case in previous years. Tobacco has seen a decrease in production in recent years and the Malawian government is looking for an alternative export crop strategy. No growth is assumed in wood energy demand for tobacco processing and the 2010 baseline value is likely to overstate demand from this industry in the coming years.

5.3.4 Results and discussion

Table 5.1 shows the estimations from equation (1) for annual per capita biomass energy consumption of different household groups in Malawi cooking with the three stone fire. The biomass needs mirror the actual food consumption quantities of different quintiles and increase with food consumed as expected. Our findings are roughly in line with per capita demand estimations of Owen et al. (2009) for Malawi and match outcomes of many other studies such as HED (2012) or Morton (2007) who find around 8–9 kg of firewood consumed per household per day in Malawi or between 291 and 574 kg per capita fuelwood consumption per year in Mali. Some regional differences are apparent in the data. Demand for cooking energy in the Northern region is often estimated to be higher than in the other two regions mainly due to the higher availability of fuelwood (Owen et al., 2009). While the latter is certainly the case, this does not necessarily have to translate into a higher consumption of firewood. If Northern Malawian diets are simply less cooking energy intensive, the demand in Northern region could be smaller than in the other regions. Our methodology of estimating demand for cooking energy directly through actual diets should thus give more accurate estimates of demand for fuelwood. The Southern region has a higher estimated per capita demand for firewood than the other two regions due to the predominance of a diet heavy in beans and rice, which are very cooking-intensive. This might be problematic since tree cover is scarcest there. In reality, the Southern dietary patterns do not result in higher consumption of firewood. Due to their supply constraints, Southern Malawians have to rely on inferior sources of fuel, especially crop residues. Fuelwood needs of households in the Central region are on average the lowest, stemming from the fact that they eat more bread, especially in urban areas, which is purchased and not baked at home. Northern households have fuelwood needs close to the Southern region as diets are similar. The numbers also show the inefficiency of charcoal production relative to firewood. Even though diets over quintiles are very similar, households cooking with charcoal require around double the amount of wood equivalent per head than households using firewood, as the majority of wood is lost during charcoal production with a conversion efficiency of 0.22.

Table 5.1: Annual per capita fuel demand per region and quintile in 2010 for three stone fire

	Firewood (kg)			Charcoal (kg wood equivalent)		
	North	Center	South	North	Center	South
Rural						
Quint 1	327.5	241.0	333.4	n.a.	n.a.	906.5
Quint 2	427.7	364.5	494.1	n.a.	n.a.	718.1
Quint 3	493.1	452.6	635.7	n.a.	828.0	1036.4
Quint 4	641.3	563.6	812.1	n.a.	821.1	1338.6
Quint 5	984.0	849.1	1180.2	1824.2	1559.9	1947.3
Urban						
Quint 1	251.5	194.7	327.3	n.a.	699.5	491.6
Quint 2	336.9	329.6	400.1	n.a.	544.8	654.0
Quint 3	368.7	381.1	417.6	907.8	855.8	864.2
Quint 4	461.3	455.5	526.3	993.1	1043.0	1006.7
Quint 5	712.3	728.3	941.8	1501.4	1455.8	1747.1

Source: Results of Malawi biomass energy demand model.

Note: Quintiles without any charcoal consumption are denoted by "n.a.".

Using household weights from the IHS we calculate initial demand for wood in 2010 considering that about 5% of each household group own an ICS as explained in the previous section. Results for initial household and non-household demand can be found in Table 5.2. We then simulate changes in household demand following population and economic growth without any change in stove preferences, i.e. the rate of technology adoption remains constant. The results in Table 5.2 show that households' fuelwood demand in all regions almost doubles from 2010 to 2030, whereas non household demand increases by about a third. Due to the high growth of household demand for biomass energy, the share of industry in biomass energy demand declines slowly from 20% in 2010 to 17% in 2030.

Table 5.2: Demand for fuelwood in Malawi in 1000 metric tons

	Initial demand (2010)	Demand projection with GDP growth and Population growth		Demand projection with GDP growth and constant population	
		2020	2030	2020	2030
Households	8,093	10,977	14,945	8,120	8,176
North	953	1,288	1,745	957	964
Center	3,082	4,179	5,692	3,093	3,117
South	4,058	5,510	7,507	4,070	4,095
Industry	2,037	2,574	3,311	n.a.	n.a.
Total	10,130	13,551	18,255	n.a.	n.a.

Source: Results of Malawi biomass energy demand model.

Note: Rate of technology adoption remains constant.

In a second step, by keeping population constant, we isolate the share of growth in household fuelwood demand due to income growth and find that in the 20 years from 2010 to 2030, a little more than 1% of demand growth originates from growth in income. This is because Malawi's GDP growth of 5% per year is simultaneously dampened by a rising population and does not translate into high income growth per capita, substantiating the assumption of no fuel switching due to income growth as explained above. The growth in charcoal use due to urbanization is quite substantial and much higher than findings of earlier studies. While the urbanization level in Malawi increases by only 3 percentage points over the 20 years in our model, our model results show that charcoal consumption more than doubles. This translates into a correspondence in which a one percentage increase in urbanization levels results in a 34% increase in charcoal demand in Malawi. Hosier et al. (1993) estimated that a one percentage increase in urbanization levels in Tanzania would lead only to a 14% increase in charcoal consumption. These high growth rates mean an enormous pressure on available supply, which is examined in the next section.

5.4 Estimating biomass supply

Our approach to estimating sustainable supply draws on existing supply estimations updated with more recent data. Similar to Owen et al. (2009) we calculate sustainable supply in the base year 2010 through assessing sustainable wood yield of different land cover classes in Malawi. While former supply estimations were derived from satellite images from the early 1990s, we use findings of a very recent land cover atlas for Malawi (FAO, 2013). Through satellite images and GIS technology the study assessed land cover change from 1990 to 2010 and gives detailed information on the areas of different land cover classes in 2010. As with household demand, we take on a regional approach when looking at the supply of biomass in Malawi. Importantly, woodfuel does not only come from forests, but also from farmland and bushland. Trees outside forests for example in the form of traditional agroforestry systems of Faidherbia albida standing on maize fields are very common in Eastern Africa (Garrity et al., 2010).

Recent data on sustainable wood yields of different land cover classes is not readily available and we rely on several different data sets which we match to the FAO land cover classes in a second step. Both Owen et al. (2009) and Millington et al. (1994) measure sustainable yields as the mean annual increment of above ground growing biomass and take care to include all woody biomass and not only the stem of trees. Owen et al. (2009) estimate sustainable yields by extrapolating data from a remote sensing inventory of indigenous trees from 1990/91 and a forest inventory from 1995/96 of Malawi. The latter was conducted during a national Biomass Supply Survey and includes estimations of annual increments of biomass growing stock (Masamba & Ngalande, 1997). Similarly, Millington et al. (1994) used a range of forest inventories and other field data estimating above ground woody biomass in Sub-Saharan Africa, which they combined with remote sensing data and extrapolated to regions where no data was available. Table 5.A3 in the appendix shows values for sustainable yields of land cover classes in Malawi from varying sources. We estimate sustainable fuelwood supply using all land cover types, but exclude protected areas such as national parks and game reserves. Forest reserves are not excluded since anecdotal evidence shows that firewood is definitely collected in forest reserves. As long as the collection is limited to sustainable yield, there is no harm in collecting branches in forest reserves. We also estimate the supply of crop residues as this an important source of biomass energy in firewood scarce regions and at harvest time (GoM, 2001). Calculating actual available biomass from crops is done through the "residue to product ratio", which estimates how the final crop output is related to the volume of residues left behind after harvest. We multiply crop output data from FAOSTAT for Malawi for 2010 with residue to product ratios from Koopmans and Koppejan (1997) to calculate the potential production of crop residues in Malawi. Still, it is not in the interest of Malawians to use crop residues as fuel, since they play an important role for food production, predominantly as fodder for livestock and to provide soil nutrients and soil organic matter for crops in the fields. While the impact of crop residue mulching on yields remains mixed depending on climate conditions (Erenstein, 2002), livestock is an important risk insurance and investment good, making crop residue fodder a considerable source of income (Dercon, 1998; Giller et al., 2009).

5.4.1 Results and discussion

The supply estimations given in Table 5.3 show that the substantial growth in fuelwood demand calculated in the previous section is faced with supply constraints depending on which data set for sustainable wood yields is used. Since it is not possible to specify the exact wood yield of a certain land cover type, our two main data sources Owen et al. (2009) and Millington et al. (1994) give minimum and maximum values for sustainable yields (see Table 5.A3). The highest numbers are based on Owen et al. (2009), who assume a rather high biomass yield compared to Millington et al. (1994) for the two largest land cover classes. Data from Owen et al. (2009) shows a yield of 0.7 ton/ha from crop land under intensive agriculture (about 3.6 million ha), which is more than three times the yield found by Millington et al. (1994). Anecdotal and visual evidence from the south of Malawi do not provide support to the hypothesis of there being many trees on-farm in the typical household. Women rather walk to the next forest reserve and bribe the soldiers protecting the reserves to let them collect firewood. Moreover, we capture crop land with trees in a separate land cover class. Similarly, Owen et al. (2009) assume an average yield of 2.5 ton/ha of wood from miombo woodland (around 3 million ha) compared to 1.3 ton/ha on average from Millington et al. (1994). The same applies to crops interspersed with trees, which can be termed as "trees outside forests". Here Millington et al. (1994) assume much lower yields compared to Owen et al. (2009). Since those trees are a significant source of fuelwood and their yield has often been ignored in the past, Millington et al. (1994) likely underestimate their yield. Overall, we expected to see higher estimated wood supply when using older yield data, since total forest cover in Malawi decreased by more than 5% in the last 20 years (FAO, 2013), but the opposite is actually the case. The actual yield numbers are likely to lie somewhere between the data provided by Owen et al. (2009) and Millington et al. (1994).

Table 5.3: Sustainable supply of fuelwood in Malawi in 2010 in 1000 metric tons

Dataset for sustainable yields	North	Center	South	National
Owen et al. (2009) Minimum	3,598	2,813	2,712	9,123
Owen et al. (2009) Maximum	5,070	3,573	3,593	12,236
Owen et al. (2009) Average	4,475	3,231	3,184	10,889
Millington et al. (1994) Minimum	1,971	967	834	3,772
Millington et al. (1994) Maximum	4,189	1,985	1,808	7,983
Millington et al. (1994) Average	3,080	1,476	1,321	5,878
Millington & Townsend (1989)	3,684	2,251	1,895	7,829
Maximum	5,465	3,693	3,720	12,878
Average	3,942	2,499	2,383	8,824
Average supply minus demand (2010)	2,989	-583	-1,675	-1,306

Source: Own calculations based on FAO (2013) and the respective datasets for sustainable yields.

Regardless of which yield data set is used, the supply estimations clearly show the imbalances between sustainable supply and demand in the different regions when compared to the initial demand in 2010 (last row of table 5.3) as already found by other studies (Owen et al., 2009). The imbalances are aggravated by the fact that not all potential sustainable supply is practically available and accessible supply might be even lower than what our numbers suggest. The Northern region has the largest supply and the smallest household demand for fuelwood, whereas the situation in the South is most alarming. The Southern excess demand could be satisfied by unsustainable harvesting of fuelwood for charcoal production, but most of this is likely to come from crop residues, while some charcoal is imported from Mozambique. Especially in the South people grow pigeon peas and use their woody stalks as fuel source. Our calculations of the potential amount of crop residues result in 8 million ton wood equivalent[3] of biomass from crop residues in Malawi, which corresponds nicely to the 12.1 million m^3 estimated by Owen et al. (2009). While crop residues hold a large potential to satisfy energy demand on a macro level, they are only available after harvest and should be a fuel of last resort due to their other uses.

For now we will use the average yield numbers of all different datasets and calculate a sustainable supply of 8.8 million tons of fuelwood, which already in 2010 shows a demand – supply gap of almost 1.5 million ton. Even the maximum estimations of sustainable supply exhibit an imbalance between demand and supply already in 2020, making it imperative to find policy measures to decrease demand and/or increase supply. Our estimations are based on land cover values for 2010 and do not encompass potential land use changes from forests into farmland following pressure from population growth.[4] This would certainly diminish supply from forests, but should be mitigated if people produce woodfuel on their own fields from agroforestry as modelled in section 5.6.

5.5 Demand side policy: improved cookstoves

A very prominent policy measure to boost both energy and food security is the dissemination of improved cookstoves. ICS can have a variety of benefits such as more efficient energy use translating into lower fuelwood requirements, as well as fewer smoke emissions potentially reducing respiratory illnesses and greenhouse gas emissions, but also higher food security through better cooked food (Lewis & Pattanayak, 2012). Since 2013 the Malawian government pursues the goal of providing ICS for 2,000,000 households by 2020. This ambitious plan would mean that almost half of Malawian households adopt ICS. In order to find out whether this policy measure will be sufficient to ensure a sustainable biomass energy sector and how less ambitious goals would affect energy demand in Malawi, we will simulate the effect of increased adoption of ICS on biomass energy demand in Malawi until 2020 and beyond.

---

3 Wood equivalent is calculated by using energy content of crop residues compared to firewood. We take the 12.5 MJ/kg energy value for air dry crop residues compared to 15 MJ/kg for firewood both with 15% moisture content from Owen et al. (2009). For a lower moisture content, the energy value for crop residues increases respectively (Wekesa, 2013).

4 Considering that forest area diminished by only 0.16% in total between 2000 and 2010 in Malawi (FAO, 2013), extreme land clearing for agriculture is unlikely to occur.

We use the estimate of 150,000 adopted ICS in 2010 from the previous section as our baseline number in 2010. The goal to provide ICS for two million households by 2020 would mean an annual compound growth rate of 30% every year[5] or that every year until 2020, 7% of the remaining households adopt ICS. We will compare this optimistic policy scenario to a pessimistic scenario where we assume an increase in ICS adoption of 1%. The percentages are evenly distributed among all household groups in all regions. To simulate the effects of ICS adoption on energy demand, it is necessary to determine how much fuelwood is actually saved compared to the three stone fire. Applying the methodology for the household baseline demand estimation in equation (1), we use specific fuel consumption values for two different stoves to calculate the fuel needs for diets of the various household groups cooking with ICS technology (Table 5.A1). The most prominent stoves distributed are the *Chitetezo Mbaula*, which is made locally from clay and relatively cheaply available, and stoves of the widely known *rocket* type, which are mainly imported into Malawi and thus not affordable for most households. Both stove types differ in their energy efficiency. Cooking tests find firewood savings for *rocket* stoves compared to the three stone fire of around 40%, while these were around 15% for the *Chitetezo* (Adkins et al., 2010a; Wang et al., 2009). In both scenarios, we will increase the number of households in Malawi using either the *Chitetezo* or *rocket* stove evenly and calculate how this changes national biomass energy demand. Population and income growth in both scenarios are identical to the demand projections in the previous section.

5.5.1 Results and discussion

In the estimation of fuelwood demand in the previous section we did not assume any change in cooking appliances until 2030. Even if the ambitious plan of the Malawian government with two million ICS adopted will not be reached, it is highly unlikely that no new ICS will be adopted. We thus compare the effect of the optimistic government's policy to a more pessimistic scenario where the group of ICS users increases by 1% every year. The results are shown in Table 5.4 and compared to the average and maximum supply estimates. Bringing two million ICS into use until 2020 does indeed reduce fuelwood demand of households and diminishes the effect of population growth on demand growth through efficiency increases. While demand growth in the pessimistic scenario closely follows population growth with a compound annual growth rate (CAGR) between 2010 and 2020 of almost 3%, the CAGR in the optimistic scenario is

5 This means that every year the group of ICS owners grows by 30% on average.

little more than half of that with 1.5%. In the optimistic scenario the group of adopters grows from 5% to 48% of Malawi's population until 2020. As the number of non-adopters decreases significantly until 2020, growth in the population share of adopters is much slower from 2020 to 2030 and increases only to 63% of the population in 2030. This effect is mirrored in the pessimistic scenario with a share of ICS adopters of 12% in 2020 and of 16% in 2030. Consequently, the substantial effect of increased ICS adoption on demand growth in the optimistic scenario slows down and population growth catches up. The CAGR of biomass energy demand in the optimistic scenario between 2020 and 2030 amounts to 2.6% and is very close to the CAGR in the pessimistic scenario of 3.1%.

Table 5.4: Demand for fuelwood in 1000 metric tons with increased improved cookstove adoption

	Initial demand (2010)	Pessimistic scenario		Optimistic scenario	
		2020	2030	2020	2030
Households	8,093	10,707	14,328	9,273	11,787
North	953	1,256	1,673	1,086	1,373
Center	3,082	4,075	5,453	3,519	4,467
South	4,058	5,376	7,203	4,669	5,947
Industry	2,037	2,574	3,311	2,574	3,311
Total	10,130	13,280	17,639	11,847	15,098
"Average supply"-demand differences	-1,306	-4,456	-8,815	-3,023	-6,274
"Maximum supply"-demand differences	2,748	-402	-4,761	1,031	-2,220

Source: Results of Malawi biomass energy demand model. Both scenarios assume GDP and population growth as in the baseline.

When comparing the scenario results for demand with the sustainable supply numbers from the previous section, the optimistic scenario is indeed able to bring total biomass demand below sustainable supply in 2020 at least for the maximum supply estimations. Even so, demand remains above supply for all the other supply estimations and is considerably above all sustainable supply estimations in 2030. Moreover, dissemination of ICS does not necessarily mean that the same number is really adopted or that less wood is used. Our pessimistic scenario might thus be more realistic, especially since the dissemination of two million ICS will require huge logistical and monetary efforts. Demand side policy measures in the form of ICS make an important contribution in decreasing biomass energy demand, but are clearly not enough to ensure a balance of demand and sustainable supply. In the next section we examine whether the latter can be ensured through supply side policy measures in the form of agroforestry.

5.6 Supply side policy: agroforestry

Research on agroforestry in Malawi began already in the 1980s under the World Agroforestry Centre, which concentrated on nitrogen-fixing tree species ("fertilizer trees") to improve soils and subsequently crop yields, mainly for the staple food maize (Garrity et al., 2010). Following promising results, an Agroforestry Food Security Programm was launched promoting and providing free seeds for several species. The predominant species was *Faidherbia albida*, a type of indigenous acacia that has been widely grown among fields for generations and can provide fuelwood and significant maize yield increases from 100–400%, albeit only after about 15 years of tree age (Garrity et al., 2010). For earlier benefits from agroforestry, other nitrogen-fixing tree shrubs for intercropping with maize were promoted: the medium term *Gliricidia sepium* and short-term species such as pigeon peas. The latter are already very common in the South of Malawi and valued both for food and firewood. While the effect of pigeon peas on maize yields is not negative but neither significantly positive due to competition in the field, *Gliricidia* has been found to increase maize yields up to threefold (Chirwa et al., 2003).

Despite positive effects of the Agroforestry Food Security Program in terms of crop yields, it was not extended after 4 years and reached only about 4% of households (Beedy et al., 2012). The reason could be that at the same time the Malawian government started its large fertilizer input subsidy program providing mineral fertilizer for maize for free and thereby suppressing the demand for fertilizer trees to increase food security. Since the subsidy program is extremely costly, the government needs alternatives to ensure food security which could come in the form of agroforestry, especially since this might establish a sustainable biomass energy sector at the same time. We thus want to simulate how a nationwide extension of agroforestry as it was promoted in the Food Security Program could look like and how this would increase the supply of biomass energy and food security.

5.6.1 Methodology

We develop a forward-looking mathematical programming model to analyze which kind of agroforestry practice farmers in Malawi are most likely to adopt given land and labor constraints. Similarly as in the Agroforestry Food Security program, households in our model have the choice between three different agroforestry options *Faidherbia*, *Gliricidia* and pigeon peas. The three types differ in fuelwood yield, effect on maize yields and labor demands as well as in the waiting period before they realize benefits for wood and maize. This baseline data and the respective sources can be found in Table 5.A4 in the annex. While *Faidherbia* does not bring any benefits for the first 15 years until maturity, ultimate benefits are much higher for this tree than for the other two shrub like trees. Households h have a finite planning horizon of 50 years (t) and maximize their sum of future utility from maize and firewood production in the form of net revenue NR_t of different agroforestry options ag subject to a discount rate and land constraints:

$$\max_{NR} U = \sum\nolimits_t NR_t * discf \tag{2}$$

Discount factors *discf* are calculated using the geometric mean of Malawi's official lending interest rate in the previous 10 years. In each period, the household decides which variable $SHARE_{ag,t,h}$ in hectares of his land under maize to allocate to each agroforestry option and derives a net revenue NR_t (equation (3)). Seeds are assumed to be provided for free by donors as was the case in the previous agroforestry program. We maintain the strong relationship between energy and food security in the way that the unit used to measure costs and benefits of agroforestry is calories/energy in mega joule per hectare (MJ/ha) depending on the area ($SHARE_{ag,t,h}$) that is allocated to each agroforestry option. This means that our model does not need any prices but costs are defined for each household by the energy requirements of the person that works on the field or collects firewood. Benefits of agroforestry thus include food calories derived from increased maize production MAI_{ag} in MJ/ha and burned calories or energy saved from not having to collect firewood. For pigeon peas, the calories of consumption of peas are also included. To convert the amount of wood produced into energy saved, the amount of time in minutes MIN_h that each household would need to collect one kg of wood is multiplied with the additional[6] energy needed for collecting wood EN_h (in MJ/minute) and the amount of wood equivalent FW_{ag} in kg/ha produced on the field. Costs are measured as energy used for work on the field, based on labor days for each agroforestry option. They include L_{ag} and LM as energy needed in MJ/ha for wood and maize production, respectively. Finally, trees can be uprooted and replanted evoking costs for uprooting CUP_{ag} and replanting $CREP_{ag}$

6 Calories burned on top of basal metabolic rate.

in MJ/ha, which are multiplied by the respective uprooted and replanted variable areas $UP_{ag,t}$ and $REP_{ag,t}$ in ha.

$$NR_t = \sum_{ag} \left(FW_{ag} * MIN_h * EN_h * SHARE_{ag,t,h} + MAI_{ag} * SHARE_{ag,t,h} - \left(L_{ag} * SHARE_{ag,t,h} \right)^{\beta} - LM * AREA_h - (CREP_{ag} * REP_{ag,t} + CUP_{ag} * UP_{ag,t}) \right) \quad (3)$$

The minutes needed by each household to collect one kg of wood MIN_h are computed by the distance $DIST_h$ to firewood collection site in minutes times the frequency of collection FRE_h divided by the weekly fuel needs in kg for each household group $FUEL_h$ from our demand model:

$$MIN_h = \frac{DIST_h}{FUEL_h} * FRE_h \quad (4)$$

Limits on land for each household group $AREA_h$ are the field sizes planted with maize, since research on agroforestry concentrated on increasing yields of maize varieties. The sum of agroforestry shares cannot exceed $AREA_h$:

$$\sum_{ag} SHARE_{ag,t,h} \leq AREA_h \quad (5)$$

Even though farm sizes are extremely small with 0.7 ha on average, labor costs are likely to matter the most when it comes to tree production on farm. Thangata et al. (2002) developed a linear programming model of improved fallow adoption based on a sample of small-scale farmers in Central Malawi. They find labor constraints to be the most important factors in determining adoption of agroforestry, which is also confirmed by Snapp et al. (2010). Moreover, Chirwa et al. (2003) find that agroforestry benefits from *Gliricidia* on farm are much lower than in the research station due to the competing labor requirements of trees and crops at the height of the rainy season. Their results go in line with the paradox explored by Alwang and Siegel (1999) that despite the small field sizes in Malawi, households' on-farm labor supply is highly constrained since households need to take on off-farm labor opportunities to overcome their lack of capital and food. Off-farm labor demand however is highest in the rainy season when labor needs on their own farms are highest as well. We therefore introduce an exponential coefficient β on labor in equation (3) to measure the effect of increased labor constraints on agroforestry choice. Moreover, by including a non-linear term, we avoid overspecialization of the household.

Data on how much firewood is needed per household type per week comes directly out of our demand model with an overall average of 47 kg per week. The IHS gives data on fuelwood collection times, but not on the number of collection trips as this differs according to location as various data for Malawi suggests.

Biran et al. (2004) found that women in the South of Malawi made the collection trip only once every 4 days collecting 53.6 kg of wood on average, since distance to collection sites was between 0.52 – 4.93 km with a mean of 241 minutes per collection trip as women had to wander through mountainous terrain. Jumbe and Angelsen (2011) on the other hand collected data with a mean of 3 collection trips per week collecting 30 kg of wood on average, while distance to collection sites in this study was only between 0.6–1.3 km. Data from the IHS shows that the majority of Malawians live less than 120 minutes away from their firewood collection site. The overall average distance to collection site is 55 minutes. We will simulate several scenarios with three different collection frequencies (1, 2 and 3) to test whether this has a significant influence on the choice of agroforestry practice.

Even though firewood is predominantly collected by women, Murphy et al. (2015) find a strong effect of gender on fuelwood source choice in Kenya as women are not allowed to practice agroforestry themselves due to cultural taboos. These cultural taboos do not exist in Malawi (Kiptot & Franzel, 2011) and Thangata et al. (2002) find that land and labor constraints are much more important for agroforestry adoption than gender. Nevertheless, we will run different scenarios for men and women to check the effect of gender on agroforestry choice by differentiating labor costs for both men and women. Both genders differ in the respective energy they use for the planting and collection activities, leading to different cost structures for women and men. In terms of land, women and man from the same household type exhibit the same land constraints. A third group of scenarios is run for two maize varieties, local and hybrid maize. Local maize varieties usually have low yields of around 1 t/ha and would benefit the most from fertilizer trees, but a lot of maize production in Malawi is done with improved hybrid seeds. Agroforestry on local maize fields might thus produce not enough wood to have a decisive effect on wood supply.

Our model allows us to calculate the potential biomass energy production in Malawi on maize fields through agroforestry given households' choice. This comes not only in the form of firewood for rural households, but we implicitly model the potential for a sustainable charcoal sector as well, since we examine a policy measure that allows to produce charcoal on crop fields through agroforestry. Charcoal produced on farm would follow the government's objective of protecting forest reserves from unsustainable harvesting. While *Gliricidia* and pigeon peas can only provide a reasonable alternative to collected firewood (their energy density is about 80% of that of firewood), both are not suitable for the production of charcoal due to their shrub-like nature. *Faidherbia albida* on the other hand can well be used for charcoal production and can thus contribute to a sustainable charcoal sector in Malawi (Barnes & Fagg, 2003).

5.6.2 Results and discussion

As explained in the previous section we run three different types of agroforestry scenarios considering frequency of firewood collection, gender and maize variety. We first examine the results for agroforestry systems combined with local maize varieties and women being the decision makers, as they are the predominant wood collectors. Since firewood collection frequency did not have any influence on choices of agroforestry systems, we do not present the results here. Land constraints and labor costs play a more important role in determining agroforestry choice than the relative benefits of wood produced on farm through energy and time saved for collection. Table 5.5 reports agroforestry choices as field shares of each agroforestry system for the third rural income quintile in each region. Other income quintiles exhibit similar and proportional regional differences. The general picture shows that all household groups prefer a mix of agroforestry systems on their fields that changes slightly over the years. While they start off with planting a larger part of their field with the two faster growing agroforestry systems that bring almost immediate benefits, the mix changes over the years in favor of *Faidherbia*. The additional energy from pigeon pea consumption does not outweigh the additional labor costs incurred, so that pigeon peas are the least-preferred agroforestry system.

Table 5.5: Agroforestry system choice as share of maize fields

Gender + Time	Region	Local Maize			Hybrid Maize		
		Faidherbia (%)	Gliricidia (%)	Pigeon Pea (%)	Faidherbia (%)	Gliricidia (%)	Pigeon Pea (%)
Female							
After 1st year	North	0.43	0.33	0.23	0.16	0.73	0.11
	Center	0.43	0.31	0.21	0.21	0.67	0.12
	South	0.34	0.43	0.23	-	0.96	0.04
After 20 years	North	0.62	0.31	0.07	0.21	0.73	0.06
	Center	0.65	0.28	0.06	0.28	0.67	0.05
	South	0.46	0.43	0.11	-	0.98	0.02

Gender + Time	Region	Local Maize			Hybrid Maize		
		Faidherbia (%)	Gliricidia (%)	Pigeon Pea (%)	Faidherbia (%)	Gliricidia (%)	Pigeon Pea (%)
Male							
After 1st year	North	0.39	0.25	0.14	0.35	0.48	0.17
	Center	0.39	0.24	0.13	0.38	0.44	0.17
	South	0.43	0.31	0.21	0.19	0.70	0.12
After 20 years	North	0.77	0.19	0.04	0.49	0.48	0.03
	Center	0.79	0.17	0.04	0.53	0.44	0.03
	South	0.65	0.28	0.06	0.25	0.70	0.05

Source: Results of Malawi agroforestry choice model.

Looking at the agroforestry mix for males shows a preference for the longer-term, high-yielding system *Faidherbia* both after 1 and after 20 years, which is much less labor intensive than the other two systems. Since men burn more energy than women, labor costs weigh more heavily in our model. This effect corresponds to the fact that off-farm labor opportunities are greater for men than for women. Some regional differences are apparent for women and men. Households in the Southern region plant more *Gliricidia* than in the other two regions. This is because field sizes in the South are significantly smaller (0.31 ha) due to a high population density, whereas field sizes are very similar in the North and Central region (0.47 and 0.51 ha, respectively). Due to the small field sizes, Southern households cannot afford to have much idle land that does not bring them any benefits in terms of higher maize yields or firewood. Similarly, this land constraint effect also determines the agroforestry system choice for hybrid maize fields in all regions, as these fields are on average 7% smaller than for local maize. The effect is starkest for Southern females who do not produce any *Faidherbia* on their small hybrid maize fields. For men the labor cost effect weighs heavier than the land constraints so that they would choose more *Faidherbia* compared to females on their hybrid maize fields, even in the South.

Since females collect firewood in Malawi and are the main producers of food crops, we use the results from the simulations for women for further calculations. Once enough firewood is produced for the household or it becomes clear that *Faidherbia* wood can well be used for charcoal production, transforming *Faidherbia* into a cash crop, men might take over agroforestry production similar as it is done in Kenya (Murphy et al., 2015). While these effects cannot be foreseen and might mean that women have to collect firewood off-farm again, an increased production of charcoal on-farm is generally a contribution to a national sustainable woodfuel sector.

Using the total amount of land under local and hybrid maize for each household group multiplied with the wood yield of their agroforestry choices, we calculate the total amount of wood produced in each year as shown in Table 5.6. Fields with local and hybrid maize in Malawi amount to more than 1 million ha or about 12 percent of total land area. Our model results show that almost 3 million tons of wood can be produced on local maize fields, the number for hybrid maize is lower with 2.4 million ton, resulting in almost 5.4 million ton national wide after 20 years, more than half of current average sustainable supply. Until 2030, demand will increase substantially as shown earlier in both cookstove scenarios. The wood production potential of agroforestry is not large enough to allow total sustainable supply to keep up with demand, even on a national level. This is not the case when sustainable supply is based on average or maximum numbers from Owen et al. (2009): These supply estimations in combination with the biomass produced through agroforestry result in sustainable supply that can satisfy demand under the optimistic ICS scenario on a national level. The results need to be further differentiated on a regional level. While the Northern region does not have a sustainable supply problem even without agroforestry interventions, agroforestry is able to ensure sufficient supply in the Central region where the ICS intervention was not enough to establish a sustainable woodfuel sector. Supply-demand imbalances remain in the Southern region, even when using maximum yield data from Owen et al. (2009). Nevertheless, sustainable supply in the Southern region almost doubles within 20 years, making households much less dependent on fuelwood collection.

Table 5.6: Fuelwood and maize produced from agroforestry on maize fields in 1000 metric tons

	Local Maize		Hybrid Maize		Total
	After 1st year	After 20 years	After 1st year	After 20 years	After 20 years
Wood					
North	48	524	27	402	926
Center	116	1,310	101	1,420	2,729
South	121	1,149	27	593	1,742
National	284	2,983	155	2,415	5,307
Surplus*	-	194	-	16	211
Maize					
National**	-	493	-	1,640	2,133

Source: Own calculations based on results of Malawi agroforestry choice model.

*Surplus after subsistence wood needs of agroforestry households are substracted.

** Note: Maize output numbers are the additional output due to agroforestry.

There is no potential for on-farm production of charcoal in the South, since households cannot cover their own wood subsistence needs. After 20 years of agroforestry, households in the Central region produce a wood surplus from *Faidherbia* of about 0.8 ton on average per year, which is very little considering that the per capita charcoal needs in a household cooking with charcoal already amount to more than 1 ton of wood equivalent per year on average (see Table 5.1). National wood surplus from agroforestry after supplying subsistence needs of agroforestry practicing households add up to 210 thousand tons per year as shown in Table 5.6. While the charcoal potential of agroforestry in Malawi is thus relatively small, potential increases in food security could be substantial. These include the indirect effect on the energy needs of wood collectors from not having to carry wood loads and the direct effect of maize yield increases. For our model we chose relatively conservative numbers on maize yield gains coming from agroforestry and the potential increase in maize output is shown in Table 5.6. The agroforestry systems chosen by households for local maize would double current national production of 470 thousand tons (FAOSTAT). The potential output gain for hybrid maize should be treated with caution. Hybrid maize yields are already relatively high due to extensive use of mineral fertilizer. It is likely that this mineral fertilizer will be substituted with fertilizer trees once their yield improving impact sets in. Thus yield gains under a hybrid maize agroforestry system are unlikely to really double production from 1688 metric ton in 2010 as our numbers would suggest. Nevertheless, the fact that agroforestry is able to replace mineral fertilizer at least partially will reduce households' expenditure and government costs for the fertilizer input subsidy. In addition there are certainly other indirect benefits from agroforestry that we are not able to calculate such as women having more time for reproductive[7] work due to the collection time saved that could contribute immensely to food security of children.

These findings demonstrate that agroforestry interventions are an essential tool to ensure sustainable supply of wood and food security in Malawi. Wood produced directly on farm is easily accessible for households and sidesteps the problem that not all of sustainable yield is practically available as mentioned above. The largest impact of agroforestry is found in the Central region in Malawi, where agroforestry can establish a sustainable biomass energy sector from an otherwise unsustainable situation. Moreover, agroforestry in Malawi can play an important role for climate change mitigation. After 20 years, the agroforestry systems produce as much wood as 1.85 million hectares of high yielding miombo woodlands. Using the average standing stock in miombo woodlands of 92 m^3/ha from Owen et al. (2009), the agroforestry systems could rescue 87,000 ha of woodlands from deforestation. Missanjo and Kamanga-Thole (2015) find a carbon stock of 30.8 tC/ha for above ground biomass of miombo woodland in Malawi, which would result into almost 10 million ton CO_2eq saved through agroforestry, making an important contribution for greenhouse gas emission savings under the REDD initiative.

5.7 Conclusion and policy implications

Biomass remains the dominant energy source in Sub-Saharan Africa especially as the main cooking fuel of rural and urban households. Conversely, traditional renewable biomass energy is missing on the agenda of the Sustainable Development Goals, even though the strong linkages to food security and the environment place biomass energy at the heart of sustainable development. Since population and GDP growth are exacerbating already existing supply-demand imbalances in highly populated countries such as Malawi, it is becoming crucial to explore policy interventions that promote sustainable biomass energy while simultaneously considering linkages with other sectors. Our study thus assessed

7 Reproductive work as opposed to productive work refers to unpaid domestic activities typically undertaken by women such as caring for and rearing children, cooking, maintenance of the household and cleaning (Beneria, 1979).

the potential for a sustainable biomass energy sector in Malawi in the light of pressures on demand and supply through population and income growth. Through a newly developed methodology we calculate demand for biomass energy based on diets and cooking habits in Malawi and project future demand from 2010 until 2030 following population and GDP growth. Sustainable supply is estimated based on wood yield data of different datasets. We simulate how the adoption of improved cookstoves could decrease biomass energy demand and develop a behavioral model to analyze the potential of widespread agroforestry adoption for promoting a sustainable biomass energy sector in Malawi.

Comparing our demand projections to sustainable supply estimations shows large supply-demand imbalances in the future, especially in the Southern region of Malawi. Even with the most optimistic assumptions of sustainable biomass yields, demand in 2030 is one third above sustainable supply. We find that demand side interventions in the form of the widespread dissemination of improved cookstoves are not enough to ensure a sustainable biomass sector in Malawi. Since the technology savings occur only once the efficient stoves are introduced, they cannot keep up with population growth. Until that time that Malawi becomes less dependent on biomass energy, supply side measures are the only way to ensure that increasing demand is met with sustainable supply. Our model results show that agroforestry systems as they are preferred by Malawian farmers have the potential to ensure sustainable fuelwood supply in 2030 even though population will almost have doubled.

Biomass is an inherently renewable resource, but it takes time for trees to grow, making it imperative to start investing in reforestation and agroforestry now. We calculate that more than 87,000 ha of woodlands can be saved from deforestation through agroforestry systems on maize fields in Malawi. At the same time, agroforestry with fertilizer trees can substitute for expensive mineral fertilizer and increase food security. Nonetheless, even with concerted policy efforts on both the demand and supply side, it is not possible to ensure sustainable biomass supply in the Southern region of Malawi due to high population growth. This emphasizes the need for a faster rural electrification and access to modern energy sources as aimed at by SDG 7[8]. At the same time the fact that traditional biomass energy remains the predominant cooking fuel in Malawi should not imply any trade-offs. As our study has shown a sustainable biomass sector in

8 SDG 7 states: "By 2030, expand infrastructure and upgrade technology for supplying modern and sustainable energy services for all in developing countries, in particular least developed countries, […]." (UN, 2015).

Malawi can be established while simultaneously protecting the environment and increasing food security.

References

Adkins, E., Tyler, E., Wang, J., Siriri, D., Modi, V., 2010a. Field testing and survey evaluation of household biomass cookstoves in rural sub-Saharan Africa. *Energy for Sustainable Development*, 14(3), 172–185.

Adkins, E., Chen, J., Winiecki, J., Koinei, P., Modi, V., 2010b. Testing institutional biomass cookstoves in rural Kenyan schools for the Millennium Villages Project. *Energy for Sustainable Development*, 14(3), 186–193.

Akinnifesi, F.K., Makumba, W., Kwesiga, F.R., 2006. Sustainable Maize Production Using Gliricidia/Maize Intercropping in Southern Malawi. *Experimental Agriculture*, 42(4), 1–17.

Alwang, J., Siegel, P.B., 1999. Labor Shortages on Small Landholdings in Malawi: Implications for Policy Reforms. *World Development*, 27(8), 1461–1475.

Bailis, R., Berrueta, V., Chengappa, C., Dutta, K., Edwards, R., Masera, O., Still, D., Smith, K.R., 2007. Performance testing for monitoring improved biomass stove interventions: experiences of the Household Energy and Health Project. *Energy for Sustainable Development*, 11(2), 57–70.

Barnes, R.D., Fagg, C.W., 2003. Faidherbia Albia, Monograph and Annotated Bibliography. Oxford Forestry Institute Department of Plant Sciences, University of Oxford, UK.

Beedy, T.L., Ajayi, O.C., Sileshi, G.W., Kundhlande, G., Chiundu, G., Simons, A.J., 2012. Scaling up Agroforestry to Achieve Food Security and Environmental Protection among Smallholder Farmers in Malawi. *Field Actions Science Reports*, Special Issue 7.

Beneria, L., 1979. Reproduction, production and the sexual division of labor. *Cambridge Journal of Economics*, 3, 203–225.

Biran, A., Abbot J., Mace, R., 2004. Families and firewood: A comparative analysis of the costs and benefits of children in firewood collection and use in two rural communities in Sub-Saharan Africa. *Human Ecology*, 32(1), 1–25.

Bunderson, W.T., Hayes, I.M., 1997. Sustainable Tobacco Production in Malawi: The Role of Wood Demand and Supply. Paper presented at the seminar on Malawi: Tobacco and the Future, Club Makokola, Mangochi, October 1997.

Chen, L., Heerink, N., Van Den Berg, M., 2006. Energy consumption in rural China: A household model for three villages in Jiangxi Province. *Ecological Economics*, 58(2), 407–420.

Chirwa, P.W., Black, C.R., Ong, C.K., Maghembe, J.A., 2003. Tree and crop productivity in gliricidia / maize / pigeonpea cropping systems in southern Malawi, *Agroforestry Systems*, 59(3), 265–277.

Choi, N., Lall, S.V., White, R., Webster, M.J., Wang, H.G., 2016. Malawi urbanization review: leveraging urbanization for national growth and development. World Bank Group, Washington, DC.

Dercon, S., 1998. Wealth, risk and activity choice: cattle in western Tanzania. *Journal of Development Economics*, 55, 1–42.

Dervis, K., J. De Melo, and S. Robinson, 1982. General Equilibrium Models for Development Policy. New York: Cambridge University Press.

Ecker, O., Qaim, M., 2011. Analyzing Nutritional Impacts of Policies: An Empirical Study for Malawi. *World Development*, 39(3), 412–428.

ECCM (Edinburgh Centre for Carbon Management), 2009. Technical Specification Reference: MOZ-TS-DIP var. Faidherbia albida. The Edinburgh Centre for Carbon Management, Edinburgh, UK.

Erenstein, O., 2002. Crop residue mulching in tropical and semi-tropical countries: an evaluation of residue availability and other technological implications. *Soil & Tillage Research*, 67, 115–133.

FAO, 2011. Human energy requirements, Report of a Joint FAO/WHO/UNU Expert Consultation Rome, 17–24 October 2001, Food and Nutrition Technical Report Series 1. Rome, Italy.

FAO, 2013. Atlas of Malawi Land Cover and Land Cover Change 1990–2010. Rome, Italy.

Garrity, D., F. Akinnifesi, O. Ajayi, S. Weldesemayat, J. Mowo, A. Kalinganire, M. Larwanou, and J. Bayala, 2010. Evergreen Agriculture: a robust approach to sustainable food security in Africa. *Food Security*, 2(3), 197–214.

Giller, K., Witter, E., Corbeels, M., Tittonell, P., 2009. Conservation agriculture and smallholder farming in Africa: The heretics' view. *Field Crops Research*, 114 (1), 23–34.

GoM (Government of Malawi), 2001. Malawi's National Forestry Programme. Lilongwe, Malawi.

Hecht, J., Kasulo, V., 2013. Development of Forest Valuation Systems, Malawi, Technical Report. Cardno Emerging Markets (UK) Ltd, London, UK.

HED, 2012. Malawi Firewood Baseline Report. Rural Malawi AMS IIG Baseline Firewood Consumption Study.

Hiemstra-van der Horst, G., Hovorka, A., 2008. Reassessing the "energy ladder": Household energy use in Maun, Botswana. *Energy Policy*, 36(9), 3333–3344.

Hosier, R.H., Mwandosya, M.J., Luhanga, M.L., 1993. Future energy development in Tanzania. The energy costs of urbanization. *Energy Policy*, 21(5), 524–542.

IEA (International Energy Agency), 2014. Africa Energy Outlook, A Focus on Energy Prospects in Sub-Saharan Africa, World Energy Outlook Special Report. Paris, France.

Jumbe, C.B.L., Angelsen, A., 2011. Modeling choice of fuelwood source among rural households in Malawi: A multinomial probit analysis. *Energy Economics*, 33(5), 732–738.

Kambewa, P., Mataya, B., Sichinga, W., Johnson, T., 2007. Charcoal: the reality – a study of charcoal consumption, trade and production in Malawi. Small and Medium Forestry Enterprise Series, 21. London: International Institute for Environment and Development.

King, R. & Byerlee, D., 1978. Factor Intensities and Locational Linkages of Rural Consumption Patterns in Sierra Leone. *American Journal of Agricultural Economics*, 60 (2), 197–206.

Kiptot, E., Franzel, S., 2011. Gender and Agroforestry in Africa: Are Women Participating?, World Agroforestry Center, Occasional Paper 13.

Koopmans, A., Koppejan, J., 1997. Agricultural and Forest Residues – Generation, Utilization and Availability. Paper presented at the Regional Consultation on Modern Applications of Biomass Energy, 6–10 January 1997, Kuala Lumpur, Malaysia.

Van der Kroon, B., Brouwer, R., van Beukering, P., 2013. The energy ladder: Theoretical myth or empirical truth? Results from a meta-analysis. *Renewable and Sustainable Energy Reviews*, 20, 504–513.

Lewis, J.J., Pattanayak, S.K., 2012. Who adopts improved fuels and cookstoves? A systematic review. *Environmental Health Perspectives*, 120(5), 637–645.

Makungwa, S., 2008. "Wood Fuel Consumption in Commercial and Institutional Sectors of Malawi" A study conducted for the development of Biomass Energy Strategy for Malawi, Bunda College of Agriculture, Lilongwe.

Malakini, M., Mwase, W., Maganga, A.M., Khonje, T., 2013. Fuelwood Use Efficiency in Cooking Technologies for Low Income Households in Malawi. *Journal of Poverty, Investment and Development*, 2, 58–63.

Masamba C.R. and Ngalande J.D., 1997. Inventory Data of Biomass Growing Stock and Supply. Centre for Social Research, University of Malawi, Zomba, Malawi.

MERA (Malawi Energy Regulatory Authority), 2016. Whither Bio and Liquefied Petroleum Gas. Available at: http://www.meramalawi.mw/index.php/

resource-center/newsletter/send/13-newsletter/34-mera-gas-pullout, access date: 04.05.2016.

Millington, A. & Townsend, J., 1989. Biomass assessment. Earthscan, London, UK.

Millington, A., Critchley, R., Douglas, T., Ryan, P., 1994. Estimating Woody Biomass in Sub-Saharan Africa. The World Bank, Washington D.C.

Minofu, M. and Kunert. M. 2014. Malawi Cookstoves Projects Survey by the MBAULA secretariat, Phase I October 2013–February 2014. Lilongwe, Malawi.

Missanjo, E., Kamanga-Thole, G., 2015. Estimation of Biomass and Carbon stock for Miombo Woodland in Dzalanyama Forest reserve, Malawi. *Research Journal of Agriculture and Forestry Sciences*, 3(3), 7–12.

Morton, J., 2007. Fuelwood Consumption and Woody Biomass accumulation in Mali, West Africa. *Ethnobotany Research & Applications*, 5, 37–44.

Murphy, D., Berazneva, J., Lee, D.R., 2015. Fuelwood Source Substitution and Shadow Prices in Western Kenya. Selected Paper prepared for presentation at the 2015 Agricultural & Applied Economics Association and Western Agricultural Economics Association Annual Meeting, San Francisco, CA, July 26–28.

Ngoma, I., Sassu, M., 2004. Sustainable African Housing through traditional Techniques and Materials. 13th World Conference on Earthquake Engineering, Vancouver, B.C., Canada August 1–6, 2004, Paper No. 170.

NSO (National Statistical Office). 2012. Integrated Household Survey 2010/2011. Lilongwe, Malawi.

Orr, B., Eiswerth, B., Finan, T., 1998. Malawi +, Public Lands Utilization Study Final Report. University of Arizona Office of Arid Lands Studies and the Forestry Research Institute of Malawi.

Owen, M., Openshaw, K., van der Plas, R., Matly, M., Hankins, M., 2009. Malawi Biomass Energy Strategy.

Owen, M., van der Plas, R., Sepp, S., 2013. Can there be energy policy in Sub-Saharan Africa without biomass?. *Energy for Sustainable Development*, 17(2), 146–152.

SMEC, 2015. National Irrigation Master Plan and Investment Framework – Main Report for Republic of Malawi, Ministry of Water Development and Irrigation, Department of Irrigation. Lilongwe, Malawi.

Snapp, S.S., Blackie, M.J., Gilbert, R.A., Bezner-Kerr, R., Kanyama-phiri, G.Y., 2010. Biodiversity can support a greener revolution in Africa. *PNAS*, 107 (48), 20840–20845.

Schuenemann, F., Thurlow, J., Zeller, M., 2016. 2017. Leveling the Field for Biofuels: Comparing the Economic and Environmental Impacts of Biofuel and Other Export Crops in Malawi. *Agricultural Economics*, 48 (3), 301–315.

Thangata, P.H., Hildebrand, P.E., Gladwin, C. H., 2002. Modeling Agroforestry Adoption and Household Decision Making in Malawi. *African Studies Quarterly*, 6 (1 & 2), 271–293.

Tschirley, D., Reardon, T., Dolislager, M., and Snyder, J., 2015. The Rise of a Middle Class in East and Southern Africa: Implications for Food System Transformation. *Journal of International Development*, 27(5), 628–646.

UN, 2015. General Assembly resolution A/70/L.1, Transforming our world: the 2030 Agenda for Sustainable Development.

UNDP, 2016. UNDP support to the implementation of sustainable development goal 7. New York, USA: United Nations Development Programme.

USAID, 2010. Evaluation of manufactured wood-burning stoves in Dadaab Refugee Camps, Kenya, USAID.

Wang, J., Tyler, E., Adkins, E., Modi, V., 2009. Field evaluation of Rocket design stoves-fuelwood use and user preferences. Ruhiira, Uganda; Mbola, Tanzania; Mwandama Malawi 2008 – 2009. Columbia University, New York.

Wekesa, A., 2013. Using GIS to assess the potential of crop residues for energy generation in Kenya. A thesis submitted in partial fulfilment of the requirements for the Degree of Master of Forestry Science in the University of Canterbury.

World Bank. 2011. Wood-Based Biomass Energy Development for Sub-Saharan Africa. Issues and Approaches. Washington DC, USA.

World Bank. 2016. World Development Indicators Online Database. Washington DC, USA.

Zulu, C.L., 2010. The forbidden fuel: Charcoal, urban woodfuel demand and supply dynamics, community forest management and woodfuel policy in Malawi. *Energy Policy*, 38(7), 3717–3730.

Annex

Table 5.A1: Specific fuel consumption by stove, fuel and food group

Food group	Specific fuel consumption					
	Wood			Charcoal**		
	3 Stone	Rocket	Chitetezo	3 Stone	Rocket	Chitetezo
Beans	11.55	7.00	9.28*	27.47	16.65	22.06
Eggs	0.50	0.20	0.26	1.18	0.48	0.62
Fish	0.50	0.20	0.26	1.18	0.48	0.62
Maize	0.97	0.45	0.71*	2.30	1.07	1.69
Meat	0.84	0.59	0.71*	1.99	1.40	1.70
Plaintains	0.65	0.35	0.50*	1.54	0.83	1.19
Poultry	0.84	0.59	0.71*	1.99	1.40	1.70
Rice	1.05	0.92*	0.79	2.49	2.18	1.88
Vegetables	0.32	0.14	0.23*	0.77	0.34	0.56

Source: Adkins et al., 2010a; Adkins et al., 2010b; Wang et al., 2009; USAID, 2010; Malakini et al., 2013.

*Values are an average between 3 Stone fire and Rocket Stove.

**The SFC of charcoal for each food group is calculated by multiplying the SFC for firewood with the ratio of the energy content of firewood (15.5 MJ/kg) to energy content of charcoal (29 MJ/kg). By dividing the SFC of charcoal through the conversion efficiency (0.22) of firewood into charcoal we get the SFC of charcoal measured in wood equivalent.

Table 5.A2: Linear expenditure system

Indices	
i,j	Commodities
h	Representative households
Parameters	
$C_{h,i}$	Consumption of commodity i by household h
P_i	Price of commodity i
Y_h	Household expenditure
$\gamma_{h,i}$	Subsistence minimum of commodity i by household h
$\beta_{h,i}$	Marginal budget share of commodity i by household h
$\epsilon_{h,i}$	Expenditure elasticity of commodity i by household h

Parameters	
$\alpha_{h,i}$	Average budget share of commodity i by household h
φ_h	Frisch parameter for household h
S_h	Supernumary income
Equations	
$C_{h,i} = P_i * \gamma_{h,i} + \beta_{h,i} * \left(Y_h - \sum_j P_j * \gamma_{h,j}\right)$	1
$\beta_{h,i} = \epsilon_{h,i} * \alpha_{h,i}$	2
$\gamma_{h,i} = \left(\frac{Y_h}{P_i}\right) * \left(\alpha_{h,i} * \frac{\beta_{h,i}}{\varphi_h}\right)$	3
$\varphi_h = \frac{Y_h}{Y_h - S_h}$	4
$S_h = \sum_i P_i * \gamma_{h,i}$	5

Source: Linear expenditure system following Dervis et al. (1982).

Table 5.A3: Sustainable yield of firewood of different land cover classes in Malawi

FAO Land cover class	Yield classification	Total land cover in ha	Sustainable yield of firewood in ton/ha								
			Owen et al. (2009) Minimum	Owen et al. (2009) Maximum	Owen et al. (2009) Average	Millington et al. (1994) Minimum	Millington et al. (1994) Maximum	Millington et al. (1994) Average	Millington & Townsend, (1989)	Maximum	Average
AGFL	Post flooding crops	400,780	0	0	0	0	0	0	0	0	0
AGHL	Crop field	91,089	0.7	0.7	0.7	0.2	0.2	0.2	0.5	0.7	0.5
AGHS	Crop field	3,591,826	0.7	0.7	0.7	0.2	0.2	0.2	0.5	0.7	0.5
AGOR	Agroforestry	7,853	2.7	3.9	3.3	0.4	2.3	1.3	2.2	3.9	2.5
AGSR	Shrubs	1,915	0.5	0.5	0.5	0.5	0.5	0.5	0.5	0.5	0.5
AGTP	Plantation	89,890	6.7	11.3	11.3	11.3	11.3	11.3	5.2	11.3	10.0
AGTR	Crops/sparse trees	568,779	2.7	3.9	3.3	0.2	0.6	0.4	0.5	3.9	1.9
ARIC	Rice	42,762	0.0	0.0	0.0	0.0	0.0	0.0	0.0	0.0	0.0
ASUG	Sugarcane	30,175	0.0	0.0	0.0	0.0	0.0	0.0	0.0	0.0	0.0
ATEA	Tea	36,944	0.0	0.0	0.0	0.0	0.0	0.0	0.0	0.0	0.0
BARE	Rock	20,711	0.0	0.0	0.0	0.0	0.0	0.0	0.0	0.0	0.0
HBCL	Savannah	522,032	0.3	0.3	0.3	0.2	0.5	0.4	0.5	0.5	0.4
HBCO	Herbaceous veg.	108,652	0.3	0.3	0.3	0.1	0.1	0.1	0.1	0.3	0.2
HBFP	Marsch	164,190	0.0	0.0	0.0	0.0	0.0	0.0	0.0	0.0	0.0
HBFT	Dambo	264,104	0.0	0.0	0.0	0.0	0.0	0.0	0.0	0.0	0.0
SRCO	Shrubland/Thicket	132,281	0.5	0.5	0.5	0.2	0.5	0.4	0.5	0.5	0.4
TREC	Evergreen	218,354	1.9	2.3	2.1	5.0	5.0	5.0	2.2	5.0	3.6
TREO	Miombo woodland	2,938,629	2.0	2.9	2.5	0.4	2.3	1.3	2.2	2.9	2.1
URBA	Houses	170,977	0.0	0.0	0.0	0.0	0.0	0.0	0.0	0.0	0.0
WAT	Water	168,167	0.0	0.0	0.0	0.0	0.0	0.0	0.0	0.0	0.0
	Total	9,570,110									

Table 5.A4: Assumptions for Malawi agroforestry system choice model

	Agroforestry system			**Source**
	Faidherbia	**Gliricidia**	**Pigeon Peas**	
Benefits				
Annual production of wood equivalent (w.e.) kg/ha				
after 1 year	-	1,209.7	2,419.4	Akinnifesi et al., 2006; Chirwa et al., 2003; ECCM, 2009.
after 6 years	-	2,822.6	2,419.4	
after 15 years	8,200.0	2,822.6	2,419.4	
*Annual production of local maize and pigeon pea ton/ha**				
after 1 year	1.0	1.0	1.0+0.5	Snapp et al., 2010; Barnes & Fagg, 2003; Chirwa et al., 2003.
after 6 years	1.0	1.9	1.0+0.5	
after 15 years	2.1	1.9	1.0+0.5	
*Energy from local maize and pigeon pea production MJ/ha per year***				
after 1 year	14,610.5	14,610.5	21,704.8	Ecker & Qaim, 2011
after 2 years	14,610.5	27,759.9	21,704.8	
after 15 years	31,266.4	27,759.9	21,704.8	
Costs				
Labor days per ha per year				
In the first year	56	58	78	SMEC, 2015
following years	0	52	78	
for maize (local/ hybird)	116/120	116/120	116/120	
Energy needed for wood production MJ/ha per year				
women				
In the first year	222.9	230.9	310.5	FAO, 2011
following years	0.0	207.0	310.5	
for maize	461.7	461.7	461.7	
men				
In the first year	289.7	300.1	403.6	FAO, 2011
following years	0.0	269.0	403.6	
for maize	600.2	600.2	600.2	

*Yields of hybrid maize are 2.7 times higher than of local maize, which usually includes mineral fertilizer. Value for pigeon peas includes maize and pigeon pea yields.

**Energy from hybrid maize is 2.7 times the energy from local maize.

6 Discussion and conclusions

Food, energy, and water are essential goods for human survival. At the same time, the three goods are intrinsically connected with each other through economic consumption and production linkages on the one hand and ecological processes on the other hand. All three are dependent on the provision of limited resources which are threatened by growing global challenges in the form of economic growth, population growth, and climate change that are particularly affecting poor developing countries. In the light of these challenges, researchers and policy makers gathered in Bonn, Germany, in 2011 and agreed that development policy cannot continue on its current "silo" path, but must undergo a transformation towards a nexus perspective of integrated food, energy, and water security policies. This study contributes twofold to the research on the food-energy-water (FEW) nexus: methodologically, through developing integrated modeling frameworks for ex-ante policy assessments that capture the linkages between food, energy, and water, and empirically through identifying those policy measures that maximize the synergies for food, energy, and water (FEW) security and minimize the tradeoffs. These empirical and methodological achievements are reviewed in the subsequent sections, followed by general policy implications and directions for further research.

6.1 Major results and achievements

6.1.1 Methodological contributions

The methodological objective of this study is the development of an integrated modeling framework for ex-ante policy assessments that encompasses the relevant linkages of the FEW nexus at country level. Due to the complexity of the nexus and the numerous linkages between food, energy, and water, there is to date no existing economy-wide modeling framework that adequately captures all various interconnections. On the one hand, economic nexus linkages are the result of market interactions around consumption and production and determine the impacts of policy measures on the economic and social spheres of the FEW nexus. These are in turn influenced by ecological processes that comprise the environmental sphere of the nexus. Natural resources such as water, soil, and biomass originate in the earth's climate system and are provided as ecosystem services by nature. The three spheres also maintain a reciprocal relationship with drivers in the form of economic and population growth as well as climate change. Global earth system models that can assess the climate system and anthropogenic feedback effects do not allow for detailed assessments of policy impacts on FEW security at country level. To overcome the disadvantages of large aggregated models and to simultaneously encompass the relevant nexus linkages, three innovative modeling approaches are developed for the policy measures assessed in Chapters 3 to 5 that concentrate on the linkages affected by each respective policy. The choice of modeling framework is guided by the three overarching research questions posed in Chapter 1:

1. What is the simultaneous impact of policy measures on food, energy, and water security?
2. What is the role of drivers in the effectiveness of policy measures?
3. How do these policies affect the livelihoods of the poorest?

A computable general equilibrium (CGE) model of Malawi is at the heart of each modeling framework. The economy-wide model not only captures all economic linkages between food, energy, and water, but also the social dimension of the FEW nexus through a detailed representation of the household sector. This is central to measure distributional effects and to answer the third research question on how policies affect the livelihoods of the poorest. A CGE model based framework is also an appropriate choice to answer the other two overarching research questions: the CGE model encompasses both the availability dimension of FEW security by measuring policy impacts on output as well as the access dimension by capturing impacts on income and prices. In addition, population and economic growth can be simulated within the CGE model to determine the role of these drivers in the effectiveness of policy measures. The CGE model approach does include neither nexus linkages outside of market interactions such as the collection of firewood for cooking energy, nor ecological processes such as the provision of water through the ecosystem that are central to FEW security. These shortcomings are overcome by linking the CGE model with different biophysical and tailor-made farm household models as suitable for each policy measure analyzed.

Chapter 3 investigates the impacts of biofuel expansion policies on FEW security. Biofuels are often blamed for displacing food crops, water resource depletion, and greenhouse gas (GHG) emissions from land clearing. These concerns underpin the "sustainability criteria" governing access to European biofuel markets. Nevertheless, it is unclear if producing biofuels in low-income countries exacerbates food insecurity or conversely increases income of smallholders. To analyze the impacts of biofuels in Malawi, a CGE model is combined with a poverty module to assess distributional effects as well as food and energy security.

Two other models are linked to the CGE model to capture the environmental dimension of the FEW nexus and the ecological linkages affected by biofuels. Water security is included by linking the CGE model of Malawi to a newly developed crop model that measures how biofuel crop expansion influences the water intensity of the Malawian agriculture. One goal of increased biofuel production is to reduce GHG emissions relative to fossil fuels to slow down global warming and climate change. The CGE model is therefore linked to an additional module that estimates changes in emissions following land use change.

In Chapter 4, a similar modeling framework is employed to analyze the effect of irrigation expansion policies on the FEW nexus. Irrigation is crucial to increase crop yields and mitigate effects from climate change, but the low profitability has led to little irrigation investments in Sub-Saharan Africa so far. As benefits from irrigation arise not only from yield increases, but also from multiplier effects and lower climate risks, the appropriate modeling framework must be able to simultaneously assess the various returns to irrigation. A CGE model and poverty module are again combined to determine the access and availability dimensions of security as well as distributional impacts. Ecological linkages are assessed by including two additional types of models. As agro-ecological conditions and crop management techniques are important drivers of the effectiveness of irrigation, a sophisticated crop model is linked to the CGE model to determine their roles in the FEW nexus. Conversely, irrigation itself is a mitigation mechanism to reduce the risk and vulnerability accruing from the driver climate change. The modeling framework is therefore extended by the inclusion of a stochastic weather simulation module to capture the linkages between uncertain climate conditions and the FEW nexus.

These two modeling frameworks however do not encompass energy security at the micro-level, which is supplied by nature as an ecosystem service, predominantly collected from woods, and not traded at markets. In Chapter 5, the impact of improved cookstoves and agroforestry on the FEW nexus is analyzed with a focus on the biomass energy sector. The role of the CGE model in this modeling framework is to assess how drivers in the form of population and economic growth affect food demand. The CGE model is then linked to a newly developed food-energy model that translates changes in food demand into changes in energy demand. This demand-side model also captures the policy impacts of increases in cooking energy efficiency through improved cookstoves. The environmental dimension and supply side of biomass energy is included through two different modeling types: a biomass supply module estimates sustainable yields of wood using different yield data sets and land cover maps, while an agroforestry farm household model simulates the impact of trees grown on farms to increase biomass supply. As a farm household model, this tailor-made mathematical programming model encompasses the social dimension of the FEW nexus by including the specific resource endowments and constraints of farm households in Malawi.

Each of these modeling frameworks does not aim to be comprehensive in a sense that all linkages between food, energy, and water are covered. Rather the frameworks specifically capture the nexus linkages affected by the policy measures analyzed to better understand the effectiveness of those policies in evoking synergies and tradeoffs. The modeling frameworks therefore close an important gap in the literature by allowing detailed policy analysis of the FEW nexus at country level, especially since all developed models can easily be transferred to other developing countries. Even if policy measures not analyzed in this study may require slightly different models, the CGE based modeling framework for the FEW nexus developed in Chapter 2 will serve as an ideal starting point. Nevertheless, the policies assessed in Chapters 3 to 5 are central nexus policies for all developing countries in a sense that they directly affect at least two nexus securities and the livelihoods of subsistence farmers, while their effectiveness is influenced by local and global drivers. The simulation modeling results on the impacts of these policy measures are examined in the next section with a view to the three guiding research questions.

6.1.2 Empirical findings

The empirical objective of this study is the identification of policy measures that maximize the synergies and minimize the tradeoffs between food, energy, and water security. Previous studies are sector-specific and unable to simultaneously capture policy effects on all three sectors. This however is vital to find those interventions that increase resource efficiencies between the three sectors and ensure the provision of food, energy, and water for all people, especially the poorest.

6.1.2.1 Simultaneous impacts on food, energy, and water security

In general, the empirical findings of the previous chapters show that policy measures always produce some tradeoffs between food, energy, and water, but that – if policies are designed correctly – these tradeoffs can be minimized while simultaneously maximizing the synergies. As expected an expansion of biofuel production from sugarcane does indeed increase macro-level energy security in Malawi by increasing the availability of fuel. Moreover, the policy simulation results of Chapter 3 cannot confirm the conception that biofuel crop expansion threatens food security, as long as sugarcane for biofuel production is irrigated. Rather food crops displace unprofitable traditional exports crops, which leads to higher food availability and lower food prices. If biofuel crops are produced by smallholders, farm households benefit from an increase in income, which increases their access to food. There is therefore a synergy between energy and food security through biofuel production policies in Malawi. This synergy is dependent on irrigation of sugarcane, which entails a tradeoff for water security. As Malawi still has enough water resources, this tradeoff diminishes in light of the synergies. Overall, there is evidence that biofuel crops are not worse but even better for the environment and food security than other export crops, which does not justify the biased reservation by policy-makers.

There is no tradeoff pertaining water security for the irrigation expansion policy analyzed in Chapter 4, since water availability and competing uses of water in Malawi are explicitly included in the irrigation potential calculations. As irrigation is conducted with smallholder technologies in the form of gravity irrigation, treadle pumps, and watering cans, energy needs are minimal. Likewise, irrigation does not require an expansion of land, which might negatively affect micro-level energy security. The policy simulation results show that a large part of irrigated land is used for food crop production, leading to higher food output and lower food prices. These findings hinge on an increase in crop yields and cropping intensity. Higher output also generates higher income for farmers, which positively affects the access dimension of FEW security. The stochastic analysis of climate variability impacts captures the stability dimension of food security. Due to the currently low irrigated crop yields in Malawi, the impact of irrigation on food stability is small but positive. If implemented correctly, irrigation expansion in Malawi is thus a policy measure that exhibits only synergies for FEW security.

When studying the impact of improved cookstoves and agroforestry policies on the nexus in Chapter 5, there is a clear focus on food and energy security. As such, water security is only indirectly – albeit positively – affected by these policies, as both lower demand for and higher supply of trees may increase the ability of the soil to hold water. The model results show that improved cookstoves and agroforestry have the potential to increase energy security in Malawi in a way that is sustainable for the biomass energy sector. Especially agroforestry increases both national availability of biomass energy as well as the access to energy as people grow their wood on their farms. As food security is crucially dependent on the secure provision of cooking energy, this increase in energy security simultaneously increases the access to and even the utilization of food, as improved cookstoves may lead to better cooked food. Agroforestry further enhances food security indirectly by increasing soil quality and crop yields. In addition, the time saved for collecting firewood through on farm production may play a crucial role for reproductive work and food security of children. The strong link between energy and food security is therefore strengthened by both interventions, leading to a win-win situation for all nexus sectors.

To conclude, all policy measures studied showed a large scope for synergies between food, energy, and water, while some tradeoffs cannot be avoided. A simultaneous increase in FEW security can be achieved through the right interventions. Nevertheless, it is important to note that in practice, the implementation of these policies will require substantial efforts, especially in terms of funding.

6.1.2.2 Role of drivers

Drivers play an important role for the success and effectiveness of all policies and are simultaneously affected by interventions. Biofuel expansion policies are promoted to increase economic growth and to reduce GHG emissions relative to fossil fuels in order to decelerate global warming. The largest economic growth potential of biofuel production in Malawi can be attained if biofuel crops are produced on plantations. But plantation production as opposed to outgrower schemes does not increase incomes of farm households, leading to a tradeoff between economic growth and food security objectives. In addition, irrigated biofuel feedstock production that is central for increases in energy and food security increases GHG emissions relative to rainfed production, which add to the already high emissions from biofuel processing. Under the current processing technology, Malawian biofuel would not meet the EU sustainability criteria. Rainfed production on the other hand, while being the most environmentally- and climate-friendly feedstock production, would entail lower food security and lower economic growth. Biofuel expansion therefore involves substantial tradeoffs that need to be valued carefully before implementing any policy measure.

Climate change is also a central driver in the expansion of irrigation. Irrigation is one of the most important climate change mitigation mechanisms and should reduce the risk of climate change impacts and the vulnerability to climate change. The results of the stochastic weather simulations in Chapter 4 indicate that irrigation expansion in Malawi is indeed able to reduce the risks from climate variability, but that at the moment the risk reduction potential is small. This is because irrigated crop yields are much lower than their potential due to inefficient input use and crop management techniques. The latter are therefore crucial drivers for the effectiveness of irrigation, especially with a view to higher food security. The importance of crop management techniques for irrigation effectiveness can also be an advantage, as these can be much easier influenced by policy makers than climate change.

Although population and GDP growth are the major drivers of increases in food demand and subsequently demand for cooking energy, the results of the modeling framework employed in Chapter 5 show that the increase in energy demand due to economic growth is almost negligible even on an economy-wide level. Conversely, demand growth due to population growth is so large that the efficiency increases from disseminating improved cookstoves are not enough to reduce demand to a sustainable level. The supply side policy of agroforestry on the other hand has the ability to provide sustainable supply even in the light of high demand growth through population growth. The nexus policy analyses of the previous chapters demonstrate how differently some drivers affect the effectiveness of policy measures and are reciprocally affected by them. Without considering the impact of drivers, policy measures might thus not be successful and entail undesirable tradeoffs.

6.1.2.3 Impact on the livelihoods of the poorest

Increases in FEW security on a national level must not translate into better access to these goods for subsistence farmers and other households living below the national poverty line. Impact of policy measures on poverty might therefore give a clearer indication on how the poorest members of society are affected and whether the access dimension of FEW security really increases for poor households.

The potential of biofuel production to decrease poverty among smallholder farmers is large, but is dependent on the production of biofuel feedstock under irrigation by smallholders themselves as opposed to plantation production. These results are also confirmed by an ex-post impact evaluation study in Malawi, which found significantly lower poverty among outgrowers of sugarcane compared to workers on sugarcane plantations (Herrmann and Grote, 2015). Farm households gain from higher income from biofuel production, which translates into higher welfare measured through consumption expenditure. In addition, net-consuming households of food benefit from lower food prices. This is particularly important for the urban poor, who also experience higher wages in the services sector as biofuel production increases demand for trade services. In the most promising biofuel policy scenario, there is a decrease in total poverty in Malawi by 1.4 percent relative to the baseline without biofuel expansion. It is important to note that lower poverty and income increases must not necessarily translate into better access to food, energy, and water. The results of Chapter 3 show that in the biofuel scenarios where a large part of land was cleared for biofuel expansion, the reduction in poverty and increases in monetary welfare are much larger compared to scenarios where biofuel production happened on already existing land. In contrast, land clearing might lead to felling of trees which would negatively affect the access to biomass energy. In this case, a decrease in poverty does not go in hand with better access to energy. Therefore, energy security is dependent on the protection of the environment.

The poverty reduction potential of irrigation is even larger compared to biofuel production, but hinges on the labor endowments of smallholder farmers and an increase in cropping intensity. Irrigation expansion as possible in Malawi considering water availability decreases poverty by more than 5 percent if farmers are not labor constrained and all irrigation infrastructure can be employed in both the rainy and dry season. Labor constraints are among the most impeding factors for adopting better crop management techniques (Woodhouse, 2009). At least in the rainy season, Malawian farmers are highly labor constrained (Wodon and Beegle, 2006). Additional labor requirements through irrigation on-farm therefore decrease their off-farm labor opportunities, leading to an increase in poverty as the results in Chapter 4 confirm. Nevertheless, lower poverty through irrigation translates into higher welfare and consumption expenditure, while the increase in food production lowers food prices for net-consuming households. Non-farm households therefore exhibit even larger poverty reductions than farm households. At the same time, energy security is not affected as an intensification of agriculture through irrigation does not require any land clearing.

In Chapter 5, there is no explicit investigation of poverty impacts of neither improved cookstove dissemination nor agroforestry. The poverty reducing effects of these policy measures rather work indirectly by increasing access and availability of biomass energy as well as the utilization dimension of food security. Similar to irrigation, agroforestry can be very labor intensive and may put additional pressure on already labor constrained farm households. This is confirmed by the farm household model results showing a clear preference for agroforestry systems that require minimal labor. As such, the farm household's labor time is not negatively affected by agroforestry. Rather the household saves time by not having to collect wood off-farm. This leaves the household time for other productive or reproductive work, which is likely to have a (monetary) poverty reducing impact. At the same time, the substantial increase in access to energy from agroforestry is clearly a decrease in "energy poverty", even though the latter term is usually used to denote a lack of modern energy sources by relying on traditional biomass energy (OECD/IEA, 2010).

Overall, the impacts of the policies analyzed on the livelihoods of poor smallholders are positive but come with many conditions, showing that the design of policy measures plays a crucial role for their success. Thereby, this study achieved its empirical objective of identifying the most beneficial policy measures for the FEW nexus and how these policies have to be designed to maximize the synergies and minimize the tradeoffs between FEW security. These findings are an essential contribution to the empirical literature through proving that even in a world with enormous pressures on limited resources, prudent policy making can provide food, energy, and water security for all. In addition, the findings emphasize the importance of integrated policy analyses that simultaneously capture impacts on different sectors and spheres to identify synergies and tradeoffs and to improve resource efficiency.

6.2 Policy implications

The findings of the previous chapters give rise to several broader policy implications for national and international development policy making. The first implication deals with the design of policies in integrating smallholders, the second with the role of funding. The third and fourth implications deal with the relationship between the FEW nexus and the SDGs as well as the unfair treatment of traditional biomass energy in national and international policy making.

6.2.1 Smallholder integration

The integration of smallholders in implementing nexus interventions proved beneficial for all policies measures analyzed: smallholder production of biofuels maximizes synergies between energy and food security; smallholder irrigation practices, which do not require any additional energy inputs, evoke synergies between water, energy, and food security; a sustainable biomass energy sector that provides secure energy and food is crucially dependent on the practice of agroforestry on smallholder maize fields. Smallholder farmers thus not only determine the economic and social spheres of the FEW nexus as producers and consumers, but have decisive influence on the environmental sphere. They directly interact with ecosystems, for example through improving the quality of soil and watershed protection by planting trees on their fields. Although many studies emphasize the importance of smallholders for simultaneously protecting the environment, achieving food security, and reducing poverty (e.g. Altieri and Koohafkan, 2008; IFAD, 2013; Riesgo et al., 2016), the actual economic growth and poverty reduction potential of smallholder farming is contested (e.g. Collier and Dercon, 2014). More than 80% of farms in Sub-Saharan Africa are smallholder farms with less than 2 hectares of land (Lowder et al., 2016), but not all smallholder farmers can move beyond subsistence production and produce profitable cash crops or practice irrigation or agroforestry (Dawson et al., 2016). The success of irrigation expansion in Malawi for example is dependent on improved crop management techniques and high labor input, which may go beyond the abilities of many family farms. Even for those smallholders that are productive enough, many challenges such as limited market access and financing options restrain their ability to manage a commercial farm (Fan et al., 2013; UNCTAD, 2015). As crucial as smallholders are for the FEW nexus, policies involving successful smallholder participation will require additional efforts in terms of removing constraints and increasing productivity of smallholders.

6.2.2 Funding of policies

Developing countries are likely unable to finance these efforts on their own. In all of the policy measures analyzed, costs for inputs and infrastructure were assumed to be financed from abroad. While the production of biofuels is lucrative for foreign investors that are already investing in sugarcane production in Malawi, irrigation expansion is not profitable enough to be undertaken without substantial investments of donors. Similarly, the cookstove and agroforestry seeds dissemination projects in Malawi are financed by NGOs or western countries' official development aid. In the end, donor support for these policy measures must simultaneously increase private sector development to sustain FEW security. Improved cookstoves for example are already produced within Malawi, albeit still dependent on the demand and distribution of NGOs. As the policy assessments have shown both biofuel and irrigation expansion in Malawi lead to growth not only in the agricultural sector, but also in the services sector and other areas of the economy, thereby strengthening the economic development of the whole country.

6.2.3 The food-energy-water nexus as a paradigm for the Sustainable Development Goals

Since the launch of the Sustainable Development Goals, national and international development policy is concerned with reaching the 17 goals, of which the FEW nexus makes up only a small portion. Similar to FEW security, the 17 SDGs form a nexus where almost all goals are interconnected with each other and pursuing one goal may lead to tradeoffs or synergies with other goals (Le Blanc, 2015; International Council for Science, 2016). This book provides two valuable lessons for the "SDG-nexus": Firstly, integrated policy analysis is necessary to identify the kind of tradeoffs and synergies that exist between different goals. Secondly, there are indeed policies that can simultaneously fulfil more than one goal and maximize synergies between different goals. At the same time, the launch of the SDGs is crucial for attaining FEW security. The SDGs go beyond pure economic growth goals and comprise the first policy agenda that integrates the social, economic, and environmental dimensions of sustainable development, which also define the FEW nexus. In this sense, the FEW nexus and the SDGs do not provide a paradigm for each other, but share a mutually beneficial relationship: Achieving FEW security simultaneously supports the SDGs and vice versa.

6.2.4 Traditional fuels for sustainable development

Finally, this study makes a contribution to advocating traditional biomass energy, as it will remain the dominating cooking fuel in Sub-Saharan Africa in the near future (IEA, 2014). In both international and national energy policies, firewood and charcoal are treated as dirty fuels that are harmful for the environment and health. The SDGs explicitly exclude "solid biofuels used for traditional purposes" from their list of sustainable modern renewable energy (UN, 2016). Charcoal remains to be officially banned by law in Malawi as in many other African countries, even though it is the main cooking fuel in the urban areas (Zulu, 2010). On the other hand, countries like Malawi and Kenya have recognized the importance of biomass energy and introduced policies for improved cookstove dissemination to reduce fuel needs and emissions from fire (Johnson et al., 2014). Nevertheless, international development policy still vilifies wood-based biomass energy as shown by the SDGs or the fact that households relying on traditional biomass energy as opposed to electricity are perceived as "energy poor" (OECD/IEA, 2010).

The research conducted in this study has shown that with the right policies, biomass energy can be inherently sustainable, improve ecosystem service availability and food security. These findings join several other studies that have been trying to absolve wood-based biomass energy from its wrongly attached unsustainable image over the years (e.g. VENRO, 2009; World Bank, 2011; Owen et al., 2013; Johnson et al., 2014). As the SDGs will determine development policy in the next 12 years, it is crucial to improve their perception of biomass energy and emphasize the importance of woodfuels for sustainable development.

6.3 Directions for further research

The development of integrated modeling frameworks is vital for quantitative analyses of policies that simultaneously affect the economic, social, and environmental spheres. As the pressure on limited natural resources is increasing, finding solutions for enhancing resource efficiencies and synergies across the three spheres requires the combination of detailed models from different disciplines, especially economics and ecology. There are different ways to combine different models: the soft link where the combined models can also be run stand-alone and independently from each other; or the hard link where the combined models are often developed simultaneously and are nested into each other through functional relationships and data (Löschel, 2006). CGE and multi-market partial equilibrium models for example have hard links between different sectors and markets. Integrated assessment models and earth system models usually feature hard links between the different spheres that they cover (Flato, 2011). Most integrated modeling frameworks with a focus on economic and social impacts on the other hand are interconnected through soft links, for example the broader IMPACT model system (Robinson et al., 2015). Similarly, the modeling frameworks developed in this study combine economic, biophysical, and farm household models through soft links. The strength of soft links is that each different model meticulously represents the structure of the sphere covered without having to compromise on how detailed economic or ecological linkages are specified. Hard links on the other hand allow for more detailed feedback loops and closer integration than soft links. The scope for more integrated models that are linked through hard links is large and will make modeling the relationships between the economy and the environment more accurate. This will make it easier to identify synergies and tradeoffs of policy measures on different parts of the economy and the environment. Yet, the establishment of hard link modeling frameworks requires comprehensive knowledge of the different models and the collaboration of researchers of different disciplines.

The use of hard links or soft links between models of the economic and environmental sphere is directly related to another area of research that will benefit from further work in the future: the integration of ecosystem services in CGE models of developing countries. The disadvantage of CGE models is their inability to capture commodities that are provided outside of markets and do not have a market price. Ecosystem services, such as firewood provided by open forests, make an important contribution to GDP in developing countries and the circular flow of income in an economy. The services include provisioning services such as water and biomass energy, regulating services such as climate regulation, cultural services such as recreation, and supporting services such as soil formation (Haines-Young and Potschin, 2009). Most of these services can only be assessed by biophysical models as they depend on numerous ecological processes and are not directly related to economic linkages. In the analyses of biofuel expansion in Chapter 3, the regulating (climate) and supporting (soil) ecosystem services were successfully soft linked to the Malawian CGE model with a GHG emissions model and a crop model. Provisioning services on the other hand are already implicitly included in the model functions of CGE models as inputs in production functions or commodities demanded by households (Bosello, 2014). Hassan and Thurlow (2011) or Calzadilla et al. (2013) for example take advantage of this CGE feature and incorporate water as an additional sector that is exogenously supplied. Similarly, environmental CGE modeling for industrial countries usually models externalities such as pollution or natural resource constraints by including them in the model equations (Wing, 2011). This hard link approach however remains a one-way link and does not cover the reciprocal feedback links between the supply of ecosystem services by nature and the demand for these services from the economy.

Finnoff and Tschirhart (2008) overcome this problem and employ an innovative modeling approach combining a CGE and an ecosystem general equilibrium model. They study the impact of different fish population recovery measures on the ecosystem and the economy, solving both models consecutively so that the results from one model serve as input to the other model and vice versa. Similarly, Robinson and Gueneau (2014) include the impact of water demand changes on the supply side by sequentially running CGE and hydrological water system models in an iterative loop. Their modeling framework of Pakistan and the Indus river basin integrates the ecosystem service water through a combination of both hard links, by including water as an economic sector and commodity in the CGE model, and soft links as water supply is modeled with an independent hydrological water basin model. The assessment of irrigation expansion in Chapter 4 employed a combination of soft and hard links as well to include climate and supporting ecosystem services in the modeling framework. An independent crop model simulates the impact of weather variability and soil quality on the growth of crops. In a second step, the impact of climate variability on crop yields is hard linked into the CGE model through changes in total factor productivity in the crop production functions. This combination of soft and hard links between models of different spheres allows the modeling framework to capture economic-environmental linkages accurately without losing important details concerning ecological processes. As such, integrating ecosystem services in economic models requires innovative and flexible modeling of economic and environmental relationships. There is still a large scope for modeling feedback effects between the economy and the environment. This study seeks to make an important contribution to integrated environmental-economic modeling of developing countries and may serve as a starting point for future research on linking the economy and the environment in models.

References

Altieri, M., Koohafkan, P., 2008. Enduring Farms: Climate Change, Smallholders and Traditional Farming Communities. Third World Network, Penang, Malaysia.

Bosello, F., 2014. Extension of the ICES CGE model with ecosystems. Research Papers Issue RP0246, Centro Euro-Mediterraneo sui Cambiamenti Climatici (CMCC), Venezia, Italy.

Collier, P., Dercon, S., 2014. African Agriculture in 50 Years: Smallholders in a Rapidly Changing World? World Development, 63, 92–101.

Dawson, N., Martin, A., Sikor, T., 2016. Green Revolution in Sub-Saharan Africa: Implications of Imposed Innovation for the Wellbeing of Rural Smallholders, *World Development*, 78, 204–218.

Fan, S., Brzeska, K., Keyzer, M., Alex Halsema, A., 2013. From Subsistence to Profit: Transforming Smallholder Farms. Food Policy Report 2013, Washington, DC: International Food Policy Research Institute.

Finnoff, D., Tschirhart, J., 2008. Linking dynamic economic and ecological general equilibrium models. Resource and Energy Economics, Vol. 30(2), 91–114.

Flato, G., 2011. Earth system models: an overview. Wiley Interdisciplinary Reviews: Climate Change, Vol. 2(6), 783–800.

Haines-Young, R., Potschin, M., 2009. The links between biodiversity, ecosystem services and human well-being. In: Raffaelli, D. and C. Frid (eds.): Ecosystem Ecology: a new synthesis. BES Ecological Reviews Series, CUP, Cambridge, UK.

Herrmann, R., Grote, U., 2015. Large-scale Agro-Industrial Investments and Rural Poverty: Evidence from Sugarcane in Malawi. Journal of African Economies, 24(5), 1–32.

IFAD, 2013. Smallholders, food security, and the environment. Rome, Italy.

Johnson, O., Sola, P., Odongo, F., Kituyi, E., Njenga, M., Iiyama, M., 2014. Sustainable energy from trees: Adopting an integrated approach to biomass energy. SEI (Stockholm Environment Institute) Policy Brief No. 20, Stockholm, Sweden.

IEA (International Energy Agency), 2014. Africa Energy Outlook, A Focus on Energy Prospects in Sub-Saharan Africa, World Energy Outlook Special Report. Paris, France.

International Council for Science, 2016. A draft framework for understanding SDG interactions. Working paper, Paris, International Council for Science (ICSU).

Le Blanc, D., 2015. Towards integration at last? The sustainable development goals as a network of targets. DESA Working Paper No. 141, ST/ESA/2015/DWP/141.

Löschel, A., 2006. Economic impact assessment by CGE models. Presentation held at "The economic modelling and integrated assessment of climate change", April 11–12, 2006, DIW Berlin, Germany.

Lowder, S.K., Skoet, J., Raney, T., 2016. The Number, Size, and Distribution of Farms, Smallholder Farms, and Family Farms Worldwide. World Development, 87, 16–29.

OECD/IEA, 2010. Energy Poverty: How to make modern energy universal? Special early excerpt of the World Energy Outlook 2010 for the UN General Assembly on the Millennium Development Goals. International Energy Agency, Paris, France.

Owen, M., van der Plas, R., Sepp, S., 2013. Can there be energy policy in Sub-Saharan Africa without biomass?. Energy for Sustainable Development, Vol. 17 (2), 146–152.

Riesgo, L., Louhichi, K., Gomez y Paloma, S., Hazell, P., Ricker-Gilbert, J., Wiggins, S., Sahn, D.E., Mishra, A., 2016. Food and nutrition security and role of smallholder farms: challenges and opportunities – Workshop proceedings. JRC Conference and Workshop Reports, European Union.

Robinson, S., Gueneau, A., 2014. Economic Evaluation of the Diamer-Basha Dam: Analysis with an Integrated Economic/Water Simulation Model of Pakistan. Pakistan Strategy Support Program Working Paper No. 14, January 2014, Washington, D.C.: International Food Policy Research Institute (IFPRI).

Robinson, S., Mason d'Croz, D., Islam, S., Sulser, T. B., Robertson, R. D., Zhu, T., Gueneau, A., Pitois, G., Rosegrant, M. W., 2015. The International Model for Policy Analysis of Agricultural Commodities and Trade (IMPACT): Model description for version 3. IFPRI Discussion Paper 1483. Washington, D.C.: International Food Policy Research Institute (IFPRI).

UNCTAD, 2015. The role of smallholder farmers in sustainable commodities production and trade. Report by the UNCTAD secretariat, Trade and Development Board (TD/B/62/9).

UN, 2016. Progress towards the Sustainable Development Goals. Report of the Secretary-General, High-level political forum on sustainable development, convened under the auspices of the Economic and Social Council (E/2016/75*).

VENRO (Association of German Development NGOs), 2009. Rethinking Biomass Energy in Sub-Sahara Africa. Bonn, Germany.

Wing, S., 2011. Computable General Equilibrium Models for the Analysis of Economy-Environment Interactions. In: Batabyal, A., Nijkamp, P., (eds.): Research Tools in Natural Resource and Environmental Economics. World Scientific Publishing Company, New Jersey, USA.

Wodon, Q., and K. Beegle. 2006. Labor Shortages Despite Underemployment? Seasonality in Time Use in Malawi. In Gender, Time Use and Poverty in Sub-Saharan Africa, Blackden, C.M., and Q. Wodon eds., 97–116. Washington DC: World Bank.

Woodhouse, P., 2009. Technology, Environment and the Productivity Problem in African Agriculture: Comment on the World Development Report 2008. Journal of Agrarian Change, Vol. 9 (2), 263–276.

World Bank. 2011. Wood-Based Biomass Energy Development for Sub-Saharan Africa. Issues and Approaches. Washington DC, USA.

Zulu, C.L., 2010. The forbidden fuel: Charcoal, urban woodfuel demand and supply dynamics, community forest management and woodfuel policy in Malawi. Energy Policy 38, 3717–3730.

General bibliography

Adkins, E., Chen, J., Winiecki, J., Koinei, P., Modi, V., 2010b. Testing institutional biomass cookstoves in rural Kenyan schools for the Millennium Villages Project. *Energy for Sustainable Development*, 14(3), 186–193.

Adkins, E., Tyler, E., Wang, J., Siriri, D., Modi, V., 2010a. Field testing and survey evaluation of household biomass cookstoves in rural sub-Saharan Africa. *Energy for Sustainable Development*, 14(3), 172–185.

Agarwal, C., Green, G.M., Grove, J.M., Evans, T.P., Schweik, C.M., 2002. A Review and Assessment of Land-Use Change Models: Dynamics of Space, Time, and Human Choice Gen. Tech. Rep. NE-297. Newtown Square, PA: U.S. Department of Agriculture, Forest Service, Northeastern Research Station.

Akinnifesi, F.K., Makumba, W., Kwesiga, F.R., 2006. Sustainable Maize Production Using Gliricidia/Maize Intercropping in Southern Malawi. *Experimental Agriculture*, 42(4), 1–17.

Aklilu, A., 2013. Water, Smallholders and Food Security – An econometric assessment of the effect of time spent on collecting water on households' economy and food security in rural Ethiopia. Master's thesis, Environmental Economics and Management Master's Programme, Faculty of Natural Resources and Agricultural Sciences, Department of Economics, Swedish University of Agricultural Sciences, Uppsala, Sweden.

Aksoy, M.A. and Isik-Dikmelik, A., 2008. Are low food prices pro-poor? Net food buyers and sellers in low income countries. World Bank, Policy Research Working Paper 4642. World Bank, Washington, DC.

Alcamo, J., et al., 2003. Ecosystems and human well-being: a framework for assessment/Millennium Ecosystem Assessment. Island Press, Washington DC, USA.

Allen, R. G., L.S. Pereira, D. Raes, Smith, M., 1998. Crop Evapotranspiration – Guidelines for Computing Crop Water Requirements. *Irrigation and Drainage paper 56*, United Nations Food and Agriculture Organization, Rome, Italy.

Altieri, M., Koohafkan, P., 2008. Enduring Farms: Climate Change, Smallholders and Traditional Farming Communities. Third World Network, Penang, Malaysia.

Alwang, J., Siegel, P.B., 1999. Labor Shortages on Small Landholdings in Malawi: Implications for Policy Reforms. *World Development*, 27(8), 1461–1475.

Armington, P., 1969. A Theory of Demand for Products Distinguished by Place of Production. Staff Papers (International Monetary Fund), 16(1), 159–178.

Arndt, C., Benfica, R., Tarp, F. Maximiano, N., Nucifora, A.M.D. & Thurlow, J., 2008. Higher Fuel and Food Prices, Economic, Impacts and Responses for Mozambique. IFPRI Discussion Paper 00836. International Food Policy Research Institute, Washington DC.

Arndt, C., Benfica, R., Tarv, F., Thurlow, J., Uaiene, R., 2010. Biofuels, poverty, and growth: a computable general equilibrium analysis of Mozambique. *Environmental and Development Economics*, 15, 81–105.

Arndt, C., Msangi, S., & Thurlow, J. 2010a. Are Biofuels Good for African Development? An Analytical Framework with Evidence from Mozambique and Tanzania. UNU-WIDER Working Paper No. 2010/110. UNU World Institute for Development Economics Research (UNU-WIDER), Helsinki, Finland.

Arndt, C., Pauw, K., Thurlow, J., 2012. Biofuels and economic development: A computable general equilibrium analysis for Tanzania. *Energy Economics*, 34(6), 1922–1930.

Arndt, C., Pauw, K., Thurlow, J., 2016. The Economy-wide Impacts and Risks of Malawi's Farm Input Subsidy Program. *American Journal of Agricultural Economics*, 98(3), 1–19.

Arora, V., 2013. An Evaluation of Macroeconomic Models for Use at EIA. U.S. Energy Information Administration Working Paper Series. U.S. Energy Information Administration, Washington DC.

Arulpragasam, J., Conway, P., 2003. Partial Equilibrium Multi-market Analysis. In Bourguignon, F., L. Pereira da Silva, L., (eds.): The Impact of Economic Policies on Poverty and Income Distribution. Oxford, UK: Oxford University Press.

Baede, A.P.M., Ahlonsou, E., Ding, Y., Schimel, D., 2001. The Climate System: An Overview. In: Climate Change 2001: The Scientific Basis. Contribution of Working Group I to the Third Assessment Report of the Intergovernmental Panel on Climate Change [Houghton, J.T., Y. Ding, D.J. Griggs, M. Noguer, P.J. van der Linden, X. Dai, K. Maskell, and C.A. Johnson (eds.)]. Cambridge University Press, Cambridge, United Kingdom and New York, NY, USA.

Bailis, R., Berrueta, V., Chengappa, C., Dutta, K., Edwards, R., Masera, O., Still, D., Smith, K.R., 2007. Performance testing for monitoring improved biomass stove interventions: experiences of the Household Energy and Health Project. *Energy for Sustainable Development*, 11(2), 57–70.

Bandyopadhyay, S., Shyamsundar, P. & Baccini, A., 2011. Forests, biomass use and poverty in Malawi. *Ecological Economics*, 70(12), 2461–2471.

Barnes, R.D., Fagg, C.W., 2003. Faidherbia Albia, Monograph and Annotated Bibliography. Oxford Forestry Institute Department of Plant Sciences, University of Oxford, UK.

Bazilian, M., Rogner, H., Howells, M., Hermann, S., Arent, D., Gielen, D., Steduto, P., Mueller, A., Komor, P., Tol, R.S.J. & Yumkella, K. K., 2011. Considering the energy, water and food nexus: Towards an integrated modelling approach. Energy Policy, 39(12), 7896–7906.

Beedy, T.L., Ajayi, O.C., Sileshi, G.W., Kundhlande, G., Chiundu, G., Simons, A.J., 2012. Scaling up Agroforestry to Achieve Food Security and Environmental Protection among Smallholder Farmers in Malawi. *Field Actions Science Reports*, Special Issue 7.

Beneria, L., 1979. Reproduction, production and the sexual division of labor. *Cambridge Journal of Economics*, 3, 203–225.

Berndes, G., 2002. Bioenergy and water: The implications of large-scale bioenergy production for water use and supply. *Global Environmental Change*, 12, 253–271.

Bernoux, M., Tinlot, M., Bockel, L., Branca, G., Gentien, A., 2011. *EX-ante carbon-balance tool (EX-ACT): technical guidelines for Version 3, Easypol module 101*. Food and Agriculture Organization of the United Nations, Rome, Italy.

Bhattacharyya, S., Timilsina, G., 2010. A review of energy system models. International Journal of Energy Sector Management, Vol. 4 (4), pp. 494–518.

Biran, A., Abbot J., Mace, R., 2004. Families and firewood: A comparative analysis of the costs and benefits of children in firewood collection and use in two rural communities in Sub-Saharan Africa. *Human Ecology*, 32(1), 1–25.

Blanco-Fonseca, M., 2010. Literature Review of Methodologies to Generate Baselines for Agriculture and Land Use. WP4 Baseline, Deliverable: D4.1, Common Agricultural Policy Regional Impact – The Rural Development Dimension Collaborative project – Small to medium-scale focused research project under the Seventh Framework Programme, Project No.: 226195, European Commission – Joint Research Centre (JRC). Sevilla, Spain.

Böhringer, C., Rutherford, T.F., Wiegard, W., 2003. Computable General Equilibrium Analysis: Opening a Black Box, Discussion Paper No. 03–56, Center for European Economic Research, Mannheim, Germany.

Bosello, F., 2014. Extension of the ICES CGE model with ecosystems. Research Papers Issue RP0246, Centro Euro-Mediterraneo sui Cambiamenti Climatici (CMCC), Venezia, Italy.

Breisinger, C., Thomas, M. & Thurlow, J., 2009. Social accounting matrices and multiplier analysis: An introduction with exercises. Food Security in Practice technical guide 5. Washington, D.C.: International Policy Research Institute.

Bunderson, W.T., Hayes, I.M., 1997. Sustainable Tobacco Production in Malawi: The Role of Wood Demand and Supply. Paper presented at the seminar on Malawi: Tobacco and the Future, Club Makokola, Mangochi, October 1997.

Burney, J.A., Naylor, R.L., 2012. Smallholder irrigation as a poverty alleviation tool in Sub-Saharan Africa. *World Development*, 40(1), 110–123.

Calzadilla, A., Rehdanz, K., Tol, R.S.J., 2011. Water scarcity and the impact of improved irrigation management: a computable general equilibrium analysis. *Agricultural Economics*, 42(3), pp. 305–323.

Calzadilla, A., Zhu, T., Rehdanz, K., Tol, R.S.J., Ringler, C., 2013. Economywide impacts of climate change on agriculture in Sub-Saharan Africa. *Ecological Economics*, 93, 150–165.

Chen, L., Heerink, N., Van Den Berg, M., 2006. Energy consumption in rural China: A household model for three villages in Jiangxi Province. *Ecological Economics*, 58, 407–420.

Chirwa, P.W., Black, C.R., Ong, C.K., Maghembe, J.A., 2003. Tree and crop productivity in gliricidia / maize / pigeonpea cropping systems in southern Malawi, *Agroforestry Systems*, 59(3), 265–277.

Choi, N., Lall, S.V., White, R., Webster, M.J., Wang, H.G., 2016. Malawi urbanization review: leveraging urbanization for national growth and development. World Bank Group, Washington, DC.

Collier, P., Dercon, S., 2014. African Agriculture in 50 Years: Smallholders in a Rapidly Changing World? World Development, 63, 92–101.

Dasgupta, S., Deichmann, U., Meisner, C., Wheeler, D., 2005. Where is the Poverty – Environment Nexus? Evidence from Cambodia, Lao PDR, and Vietnam. *World Development*, 33 (4), 617–638.

Dawson, N., Martin, A., Sikor, T., 2016. Green Revolution in Sub-Saharan Africa: Implications of Imposed Innovation for the Wellbeing of Rural Smallholders. World Development, 78, 204–218.

De Fraiture, C., Giordano, M., Liao, Y., 2008. Biofuels and implications for agricultural water use: blue impacts of green energy. *Water Policy*, 10(S1), 67–81.

Dercon, S., 1998. Wealth, risk and activity choice: cattle in western Tanzania. *Journal of Development Economics*, 55, 1–42.

Dervis, K., J. De Melo, and S. Robinson, 1982. General Equilibrium Models for Development Policy. New York: Cambridge University Press.

Diao, X., Kennedy, A., 2016. Economywide Impact of Maize Export Bans on Agricultural Growth and Household Welfare in Tanzania: A Dynamic Computable General Equilibrium Model Analysis. *Development Policy Review*, 34(1), 101–134.

Diao, X., Thurlow, J., 2012. A Recursive Dynamic Computable General Equilibrium Model, in X. Diao, J. Thurlow, S. Benin and S. Fan, eds., *Strategies and Priorities for African Agriculture: Economywide Perspectives from Country*

Studies. International Food Policy Research Institute (IFPRI), Washington, DC.

Dillon, A., 2011. The Effect of Irrigation on Poverty Reduction, Asset Accumulation, and Informal Insurance: Evidence from Northern Mali. *World Development*, 39(12), 2165–2175.

Dimaranan, Betina V., Editor (2006). Global Trade, Assistance, and Production: The GTAP 6 Data Base, Center for Global Trade Analysis, Purdue University. West Lafayette, Indiana.

Dixon, P.B., Lee, B., Muehlenbeck, T., Rimmer, M.T., Rose, A. and Vrikios, G., 2010. Effects on the U.S. of an H1N1 Epidemic: Analysis with a Quarterly CGE Model, Journal of Homeland Security and Emergency Management, 7 (1), 1–17.

Doorenbos, J., Kassam, A.H., 1979. Yield Response to water. Irrigation and Drainage paper 33, United Nations Food and Agriculture Organization, Rome, Italy.

Drechsel, P., Heffer, P., Magen, H., Mikkelsen, R., Wichelns, D. (Eds.) 2015. Managing Water and Fertilizer for Sustainable Agricultural Intensification. International Fertilizer Industry Association (IFA), International Water Management Institute (IWMI), International Plant Nutrition Institute (IPNI), and International Potash Institute (IPI). First edition, Paris, France.

Dudu, H., Chumi, S., 2008. Economics of Irrigation Water Management: A Literature Survey with Focus on Partial and General Equilibrium Models. World Bank Policy Research Working Paper 4556. Washington, DC: World Bank.

Dunkelberg, E., Finkbeiner, M., Hirschl, B., 2013. Sugarcane ethanol production in Malawi: Measures to optimize the carbon footprint and to avoid indirect emissions. *Biomass and Bioenergy*, 71, 37–45.

Easterly, W., 2015. The SDGs Should Stand for Senseless, Dreamy, Garbled. Foreign Policy, September 28, 2015. http://foreignpolicy.com/2015/09/28/the-sdgs-are-utopian-and-worthless-mdgs-development-rise-of-the-rest/, access date: 30.07.2016.

EC (European Commission), 2010. Communication from the Commission on the practical implementation of the EU biofuels and bioliquids sustainability scheme and on counting rules for biofuels. *Official Journal of the European Union*, C160, 8–16.

ECCM (Edinburgh Centre for Carbon Management), 2009. Technical Specification Reference: MOZ-TS-DIP var. Faidherbia albida. The Edinburgh Centre for Carbon Management, Edinburgh, UK.

Ecker, O., Breisinger, C. & Pauw, K., 2011: Growth is good but is not enough to improve nutrition. 2020 Conference: Leveraging Agriculture for Improving

Nutrition and Health, February 10-12, 2011, New Delhi, India. 2020 Conference Paper 7.

Ecker, O., Qaim, M., 2011. Analyzing Nutritional Impacts of Policies: An Empirical Study for Malawi. *World Development*, 39(3), 412–428.

Edwards, P., 1996. Global Comprehensive Models in Politics and Policymaking, Editorial Essay. *Climate Change*, 32, 149–161.

Elliott, J., Deryng, D., Müller, C., Frieler, K., Konzmann, M., Gerten, D., et al., 2014. Constraints and potentials of future irrigation water availability on agricultural production under climate change. In: Proceedings of the National Academy of Sciences (PNAS), 111 (9), 3239–3244.

Erenstein, O., 2002. Crop residue mulching in tropical and semi-tropical countries: an evaluation of residue availability and other technological implications. *Soil & Tillage Research*, 67, 115–133.

Ewing, M., Msangi, S., 2009. Biofuels production in developing countries: assessing tradeoffs in welfare and food security. *Environmental Science and Policy*, 12, 520–528.

Fan, S., Brzeska, K., Keyzer, M., Alex Halsema, A., 2013. From Subsistence to Profit: Transforming Smallholder Farms. Food Policy Report 2013, Washington, DC: International Food Policy Research Institute.

Fan, S., Polman, P., 2014. An ambitious development goal: Ending hunger and undernutrition by 2025. In: 2013 Global food policy report. Eds. Marble, A. and Fritschel, H., Chapter 2. Pp. 15–28. Washington, D.C.: International Food Policy Research Institute (IFPRI).

FAO, 2006. Demand for products of irrigated agriculture in sub-Saharan Africa. FAO Water Reports, 31, Food and Agriculture Organization of the United Nations, Rome, Italy.

FAO, 2008. An Introduction to the Basic Concepts of Food Security. Food Security Information for Action, Practical Guides. Published by the EC - FAO Food Security Programme, Rome, Italy.

FAO, 2011. Human energy requirements, Report of a Joint FAO/WHO/UNU Expert Consultation Rome, 17–24 October 2001, Food and Nutrition Technical Report Series 1. Food and Agriculture Organization of the United Nations, Rome, Italy.

FAO, 2013. Atlas of Malawi Land Cover and Land Cover Change 1990–2010. Food and Agriculture Organization of the United Nations, Rome, Italy.

FAO. 2015. FAOSTAT Online Database. Rome, Italy: Food and Agriculture Organization. Available at www.faostat3.org – last accessed 30 April 2015.

Fargione, F., Hill, J., Tilman, D., Polasky, S., Hawthorne, P., 2008. Land clearing and the biofuel carbon debt. *Science*, 319, 1235–1238.

Feder, G., Just, R., Zilberman, D., 1985. Adoption of Agricultural Innovations in Developing Countries: A Survey. Econ. Devel. Cult. Change, 33 (2), 255–298.

Felderer, B. & Homburg, S., 2005. Makroökonomik und neue Makroökonomik, Springer: Berlin, Heidelberg.

Finnoff, D., Tschirhart, J., 2008. Linking dynamic economic and ecological general equilibrium models. Resource and Energy Economics, Vol. 30(2), 91–114.

Fisher, M., 2004. Household welfare and forest dependence in Southern Malawi. *Environment and Development Economics*, 9, 135–154.

Flammini, A., Puri, M., Pluschke, L., Dubois, O., 2014. Walking the Nexus Talk: Assessing the Water-Energy-Food Nexus in the Context of the Sustainable Energy for All Initiative. Environment and Natural Resources Working Paper No. 58, FAO, Rome, Italy.

Flato, G., 2011. Earth system models: an overview. *Wiley Interdisciplinary Reviews: Climate Change*, 2(6), 783–800.

Garrity, D., F. Akinnifesi, O. Ajayi, S. Weldesemayat, J. Mowo, A. Kalinganire, M. Larwanou, and J. Bayala, 2010. Evergreen Agriculture: a robust approach to sustainable food security in Africa. *Food Security*, 2 (3), 197–214.

Gerbens-Leenes, P.W., Hoekstra, A.Y., 2009. The Water Footprint of Sweeteners and Bio-Ethanol from Sugar Cane, Sugar Beet and Maize. Value of Water Research Report Series No. 38, UNESCO-IHE Institute for Water Education, Delft, the Netherlands.

Giller, K., Witter, E., Corbeels, M., Tittonell, P., 2009. Conservation agriculture and smallholder farming in Africa: The heretics' view. *Field Crops Research*, 114 (1), 23–34.

GoM (Government of Malawi), 2001. Malawi's National Forestry Programme. Lilongwe, Malawi.

GOM (Government of Malawi), 2012. National Export Strategy (NES) 2013–2018, Volume 2, Annexes 1–5, Lilongwe, Malawi.

Hahn, F., 1980. General Equilibrium Theory. *The Public Interest*, 58 (special issue), 123–138.

Haines-Young, R., Potschin, M., 2009. The links between biodiversity, ecosystem services and human well-being. In: Raffaelli, D. and C. Frid (eds.): Ecosystem Ecology: a new synthesis. BES Ecological Reviews Series, CUP, Cambridge, UK.

Hassan, R., & Thurlow, J. (2011). Macro-micro feedback links of water management in South Africa: CGE analyses of selected policy regimes. *Agricultural Economics*, 42(2), 235–247.

Hecht, J., Kasulo, V., 2013. Development of Forest Valuation Systems, Malawi, Technical Report. Cardno Emerging Markets (UK) Ltd, London, UK.

HED, 2012. Malawi Firewood Baseline Report. Rural Malawi AMS IIG Baseline Firewood Consumption Study.

Herrmann, R., Grote, U., 2015. Large-scale Agro-Industrial Investments and Rural Poverty: Evidence from Sugarcane in Malawi. *Journal of African Economies*, 24(5), 1–32.

Hiemstra-van der Horst, G., Hovorka, A., 2008. Reassessing the "energy ladder": Household energy use in Maun, Botswana. *Energy Policy*, 36(9), 3333–3344.

Hoff, H. 2011. Understanding the Nexus. Background Paper for the Bonn 2011 Conference: The Water, Energy and Food Security Nexus. Stockholm, Sweden: Stockholm Environment Institute.

Hosier, R.H., Mwandosya, M.J., Luhanga, M.L., 1993. Future energy development in Tanzania. The energy costs of urbanization. *Energy Policy*, 21(5), 524–542.

Hussain, I., Hanjra, M.A., 2004. Irrigation and Poverty Alleviation: Review of the Empirical Evidence. *Irrigation and Drainage*, 53, 1–15.

ICSU, ISSC, 2015. Review of the Sustainable Development Goals: The Science Perspective. Paris: International Council for Science (ICSU). Paris, France.

IEA (International Energy Agency), 2013. Topic: Energy Security. http://www.iea.org/topics/energysecurity/, access date: 10-09-2013.

IEA (International Energy Agency), 2014. Africa Energy Outlook, A Focus on Energy Prospects in Sub-Saharan Africa, World Energy Outlook Special Report. Paris, France.

IEA (International Energy Agency), 2015. World Energy Outlook 2015. Paris, France.

IEA-ETSAP, 2014. MARKAL. http://www.iea-etsap.org/web/Markal.asp, access date: 10-01-2014.

IFAD, 2013. Smallholders, food security, and the environment. Rome, Italy.

IIASA, 2014. MESSAGE. http://www.iiasa.ac.at/web/home/research/researchPrograms/Energy/MESSAGE.en.html, access date: 10-01-2014.

Inocencio, A.; Kikuchi, M.; Tonosaki, M.; Maruyama, A.; Merrey, D.; Sally, H.; de Jong, I. 2007. Costs and performance of irrigation projects: A comparison of sub-Saharan Africa and other developing regions. Colombo, Sri Lanka: International Water Management Institute. 81 pp. (IWMI Research Report 109).

International Council for Science, 2016. A draft framework for understanding SDG interactions. Working paper, Paris, International Council for Science (ICSU).

Isaksson, A., Ng Hee, T., Robyn, G., 2005. Productivity in Developing Countries: Trends and Policies. UNIDO Research Programme, United Nations Industrial Development Organization, Vienna, Austria.

IWMI, 2007. Water for Food, Water for Life: A Comprehensive Assessment of Water Management in Agriculture. London: Earthscan, and Colombo: International Water Management Institute.

Johnson, O., Sola, P., Odongo, F., Kituyi, E., Njenga, M., Iiyama, M., 2014. Sustainable energy from trees: Adopting an integrated approach to biomass energy. SEI (Stockholm Environment Institute) Policy Brief No. 20, Stockholm, Sweden.

Jones J.W., G. Hoogenboom, C.H. Porter, K.J. Boote, W.D. Batchelor, L.A. Hunt, P.W. Wilkens, U. Singh, A.J. Gijsman, and J.T. Ritchie. 2003. The DSSAT cropping system model. European Journal of Agronomy, 18(3–4): 235–265.

Jumbe, C.B.L., Angelsen, A., 2011. Modeling choice of fuelwood source among rural households in Malawi: A multinomial probit analysis. *Energy Economics*, 33(5), 732–738.

Kambewa, P.S., Mataya, B.F., Sichinga, W.K. & Johnson, T.R., 2007. Charcoal: the reality – a study of charcoal consumption, trade and production in Malawi. Small and Medium Forestry Enterprise Series, 21. London: International Institute for Environment and Development.

Kassam, A., Lutaladio, N., Friedrich, T., Kueneman, E., Salvatore, M., Bloise, M., Tschirley, J., 2012. Natural Resource Assessment for Crop and Land Suitability: An application for selected bioenergy crops in Southern Africa region. *Integrated Crop Management*, Vol. 14, FAO, Rome, Italy.

King, C., Webber, M., 2008. Water Intensity of Transportation, *Environmental Science & Technology*, 42 (21), 7866–7872.

King, R. & Byerlee, D., 1978. Factor Intensities and Locational Linkages of Rural Consumption Patterns in Sierra Leone. *American Journal of Agricultural Economics*, 60 (2), 197–206.

Kiptot, E., Franzel, S., 2011. Gender and Agroforestry in Africa: Are Women Participating?, World Agroforestry Center, Occasional Paper 13.

Koopmans, A., Koppejan, J., 1997. Agricultural and Forest Residues – Generation, Utilization and Availability. Paper presented at the Regional Consultation on Modern Applications of Biomass Energy, 6–10 January 1997, Kuala Lumpur, Malaysia.

Kossoy, A., Peszko, G., Oppermann, K., Prytz, N., Klein, N., Blok, K., Lam, L.,Wong, L., Borkent, B., 2015. State and Trends of Carbon Pricing 2015 (September). The World Bank, Washington, DC, USA.

Le Blanc, D., 2015. Towards integration at last? The sustainable development goals as a network of targets. DESA Working Paper No. 141, ST/ESA/2015/DWP/141.

Lewis, J.J., Pattanayak, S.K., 2012. Who adopts improved fuels and cookstoves? A systematic review. *Environmental Health Perspectives*, 120 (5), 637–645.

Lofgren H., R.L. Harris and S. Robinson. 2002. A Standard Computable General Equilibrium (CGE) Model in GAMS. Washington D.C.: IFPRI.

Löschel, A., 2006. Economic impact assessment by CGE models. Presentation held at "The economic modelling and integrated assessment of climate change", April 11–12, 2006, DIW Berlin, Germany.

Lowder, S.K., Skoet, J., Raney, T., 2016. The Number, Size, and Distribution of Farms, Smallholder Farms, and Family Farms Worldwide. World Development, 87, 16–29.

Makungwa, S., 2008. "Wood Fuel Consumption in Commercial and Institutional Sectors of Malawi" A study conducted for the development of Biomass Energy Strategy for Malawi, Bunda College of Agriculture, Lilongwe.

Malakini, M., Mwase, W., Maganga, A.M., Khonje, T., 2013. Fuelwood Use Efficiency in Cooking Technologies for Low Income Households in Malawi. *Journal of Poverty, Investment and Development*, 2, 58–63.

Malmer, A., Scott, D., Vignola, R., Xu, J., 2010. Forest cover and global water governance. In: Forests and Society – Responding to Global Drivers of Change (eds Mery, G et al.), pp. 75–93. IUFRO World series, No. 25, International Union of Forest Researchers Organization (IUFRO), Vienna.

Masamba C.R. and Ngalande J.D., 1997. Inventory Data of Biomass Growing Stock and Supply. Centre for Social Research, University of Malawi. Zomba, Malawi.

McKitrick, R.R., 1998. The econometric critique of computable general equilibrium modeling: the role of functional forms. *Economic Modelling*, 15, 543–573.

Meadows, D. H., Meadows, D. L., Randers, J., Behrens III, W. W., 1972. The Limits to Growth: a report for the Club of Rome's project on the predicament of mankind. Potomac Associates, Washington DC, USA.

Mekonnen, D., Bryan, E., Alemu, T., Ringler, C., 2015. Food versus Fuel: Examining Tradeoffs in the Allocation of Biomass Energy Sources to Domestic and Productive Uses in Ethiopia. Selected Paper prepared for presentation at the 2015 Agricultural & Applied Economics Association and Western Agricultural Economics Association Annual Meeting, San Francisco, CA, July 26-28.

MERA (Malawi Energy Regulatory Authority), 2016. Whither Bio and Liquefied Petroleum Gas. Available at: http://www.meramalawi.mw/index.php/

resource-center/newsletter/send/13-newsletter/34-mera-gas-pullout, access date: 04.05.2016.

Messner, S., Schrattenholzer, L., 2000. MESSAGE – MACRO: linking an energy supply model with a macroeconomic module and solving it iteratively. *Energy*, 25, 267–282.

Millington, A. & Townsend, J., 1989. Biomass assessment. Earthscan, London, UK.

Millington, A., Critchley, R., Douglas, T., Ryan, P., 1994. Estimating Woody Biomass in Sub-Saharan Africa. The World Bank, Washington D.C.

Minofu, M. and Kunert. M. 2014. Malawi Cookstoves Projects Survey by the MBAULA secretariat, Phase I October 2013–February 2014. Lilongwe, Malawi.

Missanjo, E., Kamanga-Thole, G., 2015. Estimation of Biomass and Carbon stock for Miombo Woodland in Dzalanyama Forest reserve, Malawi. *Research Journal of Agriculture and Forestry Sciences*, 3(3), 7–12.

Mitchell, D., 2011. Biofuels in Africa: opportunities, prospects, and challenges. The World Bank, Washington, DC, USA.

Morton, J., 2007. Fuelwood Consumption and Woody Biomass accumulation in Mali, West Africa. *Ethnobotany Research & Applications*, 5, 37–44.

Mouratiadou, I., Biewald, A., Pehl, M., Bonsch, M., Baumstark, L., Klein, D., Popp, A., Luderer, G., Kriegler, E., 2016. The impact of climate change mitigation on water demand for energy and food: An integrated analysis based on the Shared Socioeconomic Pathways, *Environmental Science and Policy*, 64, 48–58.

Msangi, S., Evans, M., 2013. Biofuels and developing economies: is the timing right? *Agricultural Economics*, 44(4–5), 501–510.

Müller, C. and R.D. Robertson. 2014. Projecting future crop productivity for global economic modeling. *Agricultural Economics,* 45(1), 37–50.

Murphy, D., Berazneva, J., Lee, D.R., 2015. Fuelwood Source Substitution and Shadow Prices in Western Kenya. Selected Paper prepared for presentation at the 2015 Agricultural & Applied Economics Association and Western Agricultural Economics Association Annual Meeting, San Francisco, CA, July 26–28.

Negash, M., Swinnen, J.F.M., 2013. Biofuels and food security: Micro-evidence from Ethiopia. *Energy Policy*, 61, 963–976.

Nelson, G. C., Rosegrant, M. W., Koo, J., Robertson, Richard D., Sulser, T., Zhu, T., Ringler, C., Msangi, S., et al., 2009. Climate change: Impact on agriculture and costs of adaptation. Food Policy Report, International Food Policy Research Institute Washington, D.C.

Ngoma, I., Sassu, M., 2004. Sustainable African Housing through traditional Techniques and Materials. 13th World Conference on Earthquake Engineering, Vancouver, B.C., Canada August 1–6, 2004, Paper No. 170.

Nielsen, T., Schuenemann, F., McNulty, E., Zeller, M., Nkonya, E., Kato, E., Meyer, S., Anderson, W., Zhu, T., Queface, A., Mapemba, L., 2015. The Food-Energy-Water Security Nexus: Definitions, Policies, and Methods in an Application to Malawi and Mozambique. Discussion Paper 01480, Washington, DC: International Food Policy Research Institute.

Nin-Pratt, A., Johnson, M., Magalhaes, E., You, L., Diao, X., Chamberlin, J., 2011. Yield Gaps and Potential Agricultural Growth in West and Central Africa. IFPRI research monograph, Washington, D.C.: International Food Policy Research Institute (IFPRI).

Njewa, E., 2012. The Road Towards Low Carbon Development Strategy in Malawi. Presentation held at UNFCCC Workshop, May 18, Bonn Climate Change Conference, Bonn, Germany.

NSO (National Statistical Office), 2012. Integrated Household Survey 2010/11. Lilongwe, Malawi.

NSO (National Statistical Office), 2012a. Statistical Yearbook 2012. Zomba, Malawi.

NSO (National Statistical Office), 2012b. Integrated Household Survey 2010/11. Lilongwe, Malawi.

NSO (National Statistical Office), 2014. Annual Economic Survey Report 2010–2011. Zomba, Malawi.

OECD/IEA, 2010. Energy Poverty: How to make modern energy universal? Special early excerpt of the World Energy Outlook 2010 for the UN General Assembly on the Millennium Development Goals. International Energy Agency, Paris, France.

Orr, B., Eiswerth, B., Finan, T., 1998. Malawi +, Public Lands Utilization Study Final Report. University of Arizona Office of Arid Lands Studies and the Forestry Research Institute of Malawi.

Owen, M., Openshaw, K., van der Plas, R., Matly, M., Hankins, M., 2009. Malawi Biomass Energy Strategy.

Owen, M., van der Plas, R., Sepp, S., 2013. Can there be energy policy in Sub-Saharan Africa without biomass?. *Energy for Sustainable Development*, 17(2), 146–152.

Pagiola, S., v. Ritter, K., Bishop, J., 2004. Assessing the Economic Value of Ecosystem Conservation. World Bank Environment Department Paper No. 101, The World Bank, Washington DC, USA.

Pattanayak, S., Sills, E., Kramer, R., 2004. Seeing the forest for the fuel. *Environment and Development Economics*, 9, 155–179.

Pauw, K., & Thurlow, J., 2011. Agricultural growth, poverty, and nutrition in Tanzania. *Food Policy*, 36(6), 795–804.

Pauw, K., Beck, U., Mussa, R., 2014. Did rapid smallholder-led agricultural growth fail to reduce rural poverty? Making sense of Malawi's poverty puzzle. WIDER Working Paper 2014/123.

Pauw, K., Schuenemann, F., Thurlow, J., 2015. A 2010 Social Accounting Matrix for Malawi. Unpublished mimeograph, International Food Policy Research Institute, Washington DC, USA.

Pauw, K., Thurlow, J., Bachu, M., van Seventer, D., 2011. The economic costs of extreme weather events: a hydrometeorological CGE analysis for Malawi. *Environment and Development Economics,* 16, 177–198.

Pauw, K., Thurlow, J., Uaiene, R. & Mazunda, J., 2012. Agricultural Growth and Poverty in Mozambique: Technical Analysis in Support of the Comprehensive Africa Agriculture Development Program (CAADP). Mozambique Strategy Support Program, Working Paper 2.

Penning de Vries, F., Acquay, H., Molden, D., Scherr, S., Valentin, C., Cofie, O., 2003. Integrated land and water management for food and environmental security. Comprehensive Assessment of Water Management in Agriculture Research Report 1. Colombo, Sri Lanka: Comprehensive Assessment Secretariat.

Piermartini, R.; Teh, R., 2005. Demystifying Modelling Methods for Trade Policy. WTO Discussion Paper 10 (September), World Trade Organization, Geneva.

Pollak, R. A., Wales, T. J., 1969. Estimation of the Linear Expenditure System. *Econometrica*, 37(4), pp. 611–628.

Ponce, R., Bosello, F., & Giupponi, C., 2012. Integrating Water Resources into Computable General Equilibrium Models – A Survey. Fondazione Eni Enrico Mattei Working Paper Series 57.2012.

Quintero, J.A., Cardona, C.A., Felix, E., Moncada, J., Sánchez, O.J., Gutiérrez, L.F., 2012. Techno-economic analysis of bioethanol production in Africa: Tanzania case. *Energy*, 48(1), 442–454.

Riesgo, L., Louhichi, K., Gomez y Paloma, S., Hazell, P., Ricker-Gilbert, J., Wiggins, S., Sahn, D.E., Mishra, A., 2016. Food and nutrition security and role of smallholder farms: challenges and opportunities – Workshop proceedings. JRC Conference and Workshop Reports, European Union.

Ringler, C., A. Bhaduri, and R. Lawford. 2013. The Nexus across Water, Energy, Land and Food (WELF): Potential for Improved Resource Use Efficiency? *Current Opinion in Environmental Sustainability*, 5, 617–624.

Robinson, S., Cattaneo, A., El-Said, M. 2001. "Updating and Estimating a Social Accounting Matrix Using Cross Entropy Methods," *Economic Systems Research*, 13(1), 47–64.

Robinson, S., Gueneau, A., 2014. Economic Evaluation of the Diamer-Basha Dam: Analysis with an Integrated Economic/Water Simulation Model of Pakistan. Pakistan Strategy Support Program Working Paper No. 14, January 2014, Washington, D.C.: International Food Policy Research Institute (IFPRI).

Robinson, S., Mason d'Croz, D., Islam, S., Sulser, T. B., Robertson, R. D., Zhu, T., Gueneau, A., Pitois, G., Rosegrant, M. W., 2015. The International Model for Policy Analysis of Agricultural Commodities and Trade (IMPACT): Model description for version 3. IFPRI Discussion Paper 1483. Washington, D.C.: International Food Policy Research Institute (IFPRI).

Rodrigues, J., Thurlow, J., Landman, W., Ringler, C., Robertson, R., Tingju, Z., 2016. The economic value of seasonal forecasts stochastic economywide analysis for East Africa. IFPRI Discussion Paper 1546. Washington, D.C.: International Food Policy Research Institute (IFPRI).

Rosegrant, M.W. and the IMPACT Development Team, 2012. International Model for Policy Analysis of Agricultural Commodities and Trade (IMPACT): Model Description. International Food Policy Research Institute (IFPRI), Washington, DC.

Rosegrant, M.W., Zhu, T., Msangi, S., Sulser, T., 2008. Global scenarios for biofuels: Impacts and implications. *Review of Agricultural Economics*, 30(3), 495–505.

Rosenzweig, C., Parry, M.L., 1994. Potential impact of climate change onworld food supply. *Nature*, 367, 133–138.

Sadoulet, E., De Janvry, A., 1995. Quantitative development policy analysis. Johns Hopkins University Press, Baltimore and London.

Schuenemann, F., Thurlow, J., Zeller, M., 2016. Leveling the Field for Biofuels: Comparing the Economic and Environmental Impacts of Biofuel and Other Export Crops in Malawi. IFPRI Discussion Paper 01500.

Schuenemann, F., Thurlow, J., Zeller, M., 2016. 2017. Leveling the Field for Biofuels: Comparing the Economic and Environmental Impacts of Biofuel and Other Export Crops in Malawi. *Agricultural Economics*, 48 (3), 301–315.

Searchinger, T., Heimlich, R., Houghton, R.A., Dong, F., Elobeid, A., Fabiosa, J., Tokgoz, S., Hayes, D., Yu, T-H., 2008. Use of U.S. Croplands for Biofuels Increases Greenhouse Gases Through Emissions from Land-Use Change. *Science*, 319(5867), 1238–1240.

SEI, 2013. LEAP brochure, http://www.sei-us.org/Publications_PDF/SEI-LEAP-brochure-Jan2012.pdf, access date: 10-01-2014.

Shoven, J., Whalley, J., 1984. Applied General-Equilibrium Models of Taxation and International Trade: An Introduction and Survey. *Journal of Economic Literature*, 22(3), 1007–1051.

Singh, I., Squire, L., Strauss, J. [editors], 1986. Agricultural household models: extensions, applications, and policy. Baltimore, MD: The Johns Hopkins University Press.

SMEC, 2015. National Irrigation Master Plan and Investment Framework – Main Report for Republic of Malawi, Ministry of Water Development and Irrigation, Department of Irrigation. Lilongwe, Malawi.

Snapp, S.S., Blackie, M.J., Gilbert, R.A., Bezner-Kerr, R., Kanyama-phiri, G.Y., 2010. Biodiversity can support a greener revolution in Africa. *PNAS*, 107 (48), 20840–20845.

Stone, R., 1954. Linear Expenditure Systems and Demand Analysis: An Application to the Pattern of British Demand. *The Economic Journal*, 64(255), 511–527.

Svendsen, M., Ewing, M., Msangi, S., 2009. Measuring Irrigation Performance in Africa. Discussion Paper 00894, Washington, DC: International Food Policy Research Institute.

Taylor, J., Adelman, I., 2003. Agricultural Household Models: Genesis, Evolution, and Extensions. *Review of Economics of the Household*, 1(1), 33–58.

Thangata, P.H., Hildebrand, P.E., Gladwin, C. H., 2002. Modeling Agroforestry Adoption and Household Decision Making in Malawi. *African Studies Quarterly*, 6 (1 & 2), 271–293.

Thurlow, J., Branca, G., Felix, E., Maltsoglou, I., Rincón, L. E., 2015. Producing Biofuels in Low-Income Countries: An Integrated Environmental and Economic Assessment for Tanzania. *Environmental and Resource Economics*, 64(2), 1–19.

Timilsina, G. R., Beghin, J. C., van der Mensbrugghe, D., Mevel, S., 2012. The impacts of biofuels targets on land-use change and food supply: A global CGE assessment. *Agricultural Economics*, 43, 315–332.

Tschirley, D., Reardon, T., Dolislager, M., and Snyder, J., 2015. The Rise of a Middle Class in East and Southern Africa: Implications for Food System Transformation. *Journal of International Development*, 27(5), 628–646.

UN, 1987. Report of the World Commission on Environment and Development: Our Common Future (Brundtland Report). United Nations, New York, USA.

UN, 2000. General Assembly resolution A/55/L.2, United Nations Millennium Declaration. United Nations, New York, USA.

UN, 2015. General Assembly resolution A/70/L.1, Transforming our world: the 2030 Agenda for Sustainable Development.

UN, 2015a. The Millennium Development Goals Report 2015. United Nations, New York, USA.

UN, 2015b. General Assembly resolution A/70/L.1, Transforming our world: the 2030 Agenda for Sustainable Development. United Nations, New York, USA.

UN, 2016. Progress towards the Sustainable Development Goals. Report of the Secretary-General, High-level political forum on sustainable development, convened under the auspices of the Economic and Social Council (E/2016/75*).

UNCTAD, 2015. The role of smallholder farmers in sustainable commodities production and trade. Report by the UNCTAD secretariat, Trade and Development Board (TD/B/62/9).

UNEP, 1999. Climate Change Mitigation Study in Southern Africa, Tanzania Country Study. UNEP Collaborating Centre on Energy and Environment, Risø National Laboratory, Denmark.

UN-Water, 2013. *Water Security & the Global Water Agenda*. http://www.unwater.org/downloads/watersecurity_analyticalbrief.pdf, access date: 10-09-2013.

USAID, 2010. Evaluation of manufactured wood-burning stoves in Dadaab Refugee Camps, Kenya, USAID.

Van der Kroon, B., Brouwer, R., van Beukering, P., 2013. The energy ladder: Theoretical myth or empirical truth? Results from a meta-analysis. *Renewable and Sustainable Energy Reviews*, 20, 504–513.

Van Wijk, M.T., Rufino, M.C., Enahoro, D., Parsons, D., Silvestri, S., Valdivia, R.O., Herrero, M., 2014. Farm household models to analyse food security in a changing climate: A review. *Global Food Security*, 3, 77–84.

VENRO (Association of German Development NGOs), 2009. Rethinking Biomass Energy in Sub-Sahara Africa. Bonn, Germany.

Wakeford, J., Kelly, C. and Mentz Lagrange, S. 2015. Mitigating risks and vulnerabilities in the energy-food-water nexus in developing countries: Summary for Policymakers. Sustainability Institute, South Africa.

Wang, J., Tyler, E., Adkins, E., Modi, V., 2009. Field evaluation of Rocket design stoves-fuelwood use and user preferences. Ruhiira, Uganda; Mbola, Tanzania; Mwandama Malawi 2008–2009. Columbia University, New York.

Watson, H. K., 2011. Potential to expand sustainable bioenergy from sugarcane in southern Africa. *Energy Policy*, 39(10), 5746–5750.

WEAP21, 2014. Water Evaluation and Planning. http://www.weap21.org/index.asp, access date: 10-01-2014.

Wekesa, A., 2013. Using GIS to assess the potential of crop residues for energy generation in Kenya. A thesis submitted in partial fulfilment of the requirements for the Degree of Master of Forestry Science in the University of Canterbury.

Welsh, M., Hermann, S., Howells, M., Rogner, H. H., Young, C., Ramma, I., Bazilian, M., Fischer, G., Alfstad, T., Gielen, D., Le Blanc, D., Röhrl, A., Steduto, P. and A. Müller, 2014. Adding value with CLEWS – Modelling the energy system and its interdependencies for Mauritius. *Applied Energy*, 113, 1434–1445.

Wing, I.S., 2004. Computable General Equilibrium Models and Their Use in Economy-Wide Policy Analysis. MIT Joint Program on the Science and Policy of Global Change, Technical Note No. 6.

Wing, S., 2011. Computable General Equilibrium Models for the Analysis of Economy-Environment Interactions. In: Batabyal, A., Nijkamp, P., (eds.): Research Tools in Natural Resource and Environmental Economics. World Scientific Publishing Company, New Jersey, USA.

Wodon, Q., and K. Beegle. 2006. Labor Shortages Despite Underemployment? Seasonality in Time Use in Malawi. In Gender, Time Use and Poverty in Sub-Saharan Africa, Blackden, C.M., and Q. Wodon eds., 97–116. Washington DC: World Bank.

Woodhouse, P., 2009. Technology, Environment and the Productivity Problem in African Agriculture: Comment on the World Development Report 2008. *Journal of Agrarian Change*, 9(2), 263–276.

World Bank, 2009. Environmental Crisis or Sustainable Development Opportunity? Transforming the charcoal sector in Tanzania. A Policy Note. Washington DC, USA.

World Bank. 2011. Wood-Based Biomass Energy Development for Sub-Saharan Africa. Issues and Approaches. Washington DC, USA.

World Bank. 2016. World Development Indicators Online Database. Washington DC, USA.

World Bank. 2017. World Development Indicators Online Database. Washington DC, USA.

You, L., Ringler, C., Wood-Sichra, U., Robertson, R., Wood, S., Zhu, T., Nelson, G., Guo, Z., Sun, Y., 2011. What is the irrigation potential for Africa? A combined biophysical and socioeconomic approach. *Food Policy*, 36(6), 770–782.

Zilberman, D., Hochman, G., Rajagopal, D., Sexton, S., Timilsina, G., 2013. The impact of biofuels on commodity food prices: Assessment of findings. *American Journal of Agricultural Economics*, 95(2), 275–281.

Zulu, C.L., 2010. The forbidden fuel: Charcoal, urban woodfuel demand and supply dynamics, community forest management and woodfuel policy in Malawi. Energy Policy 38, 3717–3730.

Development Economics and Policy

Series edited by Franz Heidhues†, Joachim von Braun, Ulrike Grote and Manfred Zeller

Vol. 1 Andrea Fadani: Agricultural Price Policy and Export and Food Production in Cameroon. A Farming Systems Analysis of Pricing Policies. The Case of Coffee-Based Farming Systems. 1999.

Vol. 2 Heike Michelsen: Auswirkungen der Währungsunion auf den Strukturanpassungsprozeß der Länder der afrikanischen Franc-Zone. 1995.

Vol. 3 Stephan Bea: Direktinvestitionen in Entwicklungsländern. Auswirkungen von Stabilisierungsmaßnahmen und Strukturreformen in Mexiko. 1995.

Vol. 4 Franz Heidhues / François Kamajou: Agricultural Policy Analysis – Proceedings of an International Seminar, held at the University of Dschang, Cameroon on May 26 and 27 1994, funded by the European Union under the Science and Technology Program (STD). 1996.

Vol. 5 Elke M. Förster: Protection or Liberalization? A Policy Analysis of the Korean Beef Sector. 1996.

Vol. 6 Gertrud Schrieder: The Role of Rural Finance for Food Security of the Poor in Cameroon. 1996.

Vol. 7 Nestor R. Ahoyo Adjovi: Economie des Systèmes de Production intégrant la Culture de Riz au Sud du Bénin: Potentialités, Contraintes et Perspectives. 1996.

Vol. 8 Jenny Müller: Income Distribution in the Agricultural Sector of Thailand. Empirical Analysis and Policy Options. 1996.

Vol. 9 Michael Brüntrup: Agricultural Price Policy and its Impact on Production, Income, Employment and the Adoption of Innovations. A Farming Systems Based Analysis of Cotton Policy in Northern Benin. 1997.

Vol. 10 Justin Bomda: Déterminants de l'Epargne et du Crédit, et leurs Implications pour le Développement du Système Financier Rural au Cameroun. 1998.

Vol. 11 John M. Msuya: Nutrition Improvement Projects in Tanzania: Implementation, Determinants of Performance, and Policy Implications. 1998.

Vol. 12 Andreas Neef: Auswirkungen von Bodenrechtswandel auf Ressourcennutzung und wirtschaftliches Verhalten von Kleinbauern in Niger und Benin. 1999.

Vol. 13 Susanna Wolf (ed.): The Future of EU-ACP Relations. 1999.

Vol. 14 Franz Heidhues / Gertrud Schrieder (eds.): Romania – Rural Finance in Transition Economies. 2000.

Vol. 15 Katinka Weinberger: Women's Participation. An Economic Analysis in Rural Chad and Pakistan. 2000.

Vol. 16 Christof Batzlen: Migration and Economic Development. Remittances and Investments in South Asia: A Case Study of Pakistan. 2000.

Vol. 17 Matin Qaim: Potential Impacts of Crop Biotechnology in Developing Countries. 2000.

Vol. 18 Jean Senahoun: Programmes d'ajustement structurel, sécurité alimentaire et durabilité agricole. Une approche d'analyse intégrée, appliquée au Bénin. 2001.

Vol. 19 Torsten Feldbrügge: Economics of Emergency Relief Management in Developing Countries. With Case Studies on Food Relief in Angola and Mozambique. 2001.

Vol. 20 Claudia Ringler: Optimal Allocation and Use of Water Resources in the Mekong River Basin: Multi-Country and Intersectoral Analyses. 2001.

Vol. 21 Arnim Kuhn: Handelskosten und regionale (Des-)Integration. Russlands Agrarmärkte in der Transformation. 2001.

Vol. 22 Ortrun Anne Gronski: Stock Markets and Economic Growth. Evidence from South Africa. 2001.

Vol. 23 Patrick Webb / Katinka Weinberger (eds.): Women Farmers. Enhancing Rights, Recognition and Productivity. 2001.

Vol. 24 Mingzhi Sheng: Lebensmittelkonsum und -konsumtrends in China. Eine empirische Analyse auf der Basis ökonometrischer Nachfragemodelle. 2002.

Vol. 25 Maria Iskandarani: Economics of Household Water Security in Jordan. 2002.

Vol. 26 Romeo Bertolini: Telecommunication Services in Sub-Saharan Africa. An Analysis of Access and Use in the Southern Volta Region in Ghana. 2002.

Vol. 27 Dietrich Müller-Falcke: Use and Impact of Information and Communication Technologies in Developing Countries' Small Businesses. Evidence from Indian Small Scale Industry. 2002.

Vol. 28 Wolfram Erhardt: Financial Markets for Small Enterprises in Urban and Rural Northern Thailand. Empirical Analysis on the Demand for and Supply of Financial Services, with Particular Emphasis on the Determinants of Credit Access and Borrower Transaction Costs. 2002.

Vol. 29 Wensheng Wang: The Impact of Information and Communication Technologies on Farm Households in China. 2002.

Vol. 30 Shyamal K. Chowdhury: Institutional and Welfare Aspects of the Provision and Use of Information and Communication Technologies in the Rural Areas of Bangladesh and Peru. 2002.

Vol. 31 Annette Luibrand: Transition in Vietnam. Impact of the Rural Reform Process on an Ethnic Minority. 2002.

Vol. 32 Felix Ankomah Asante: Economic Analysis of Decentralisation in Rural Ghana. 2003.

Vol. 33 Chodechai Suwanaporn: Determinants of Bank Lending in Thailand: An Empirical Examination for the Years 1992 to 1996. 2003.

Vol. 34 Abay Asfaw: Costs of Illness, Demand for Medical Care, and the Prospect of Community Health Insurance Schemes in the Rural Areas of Ethiopia. 2003.

Vol. 35 Gi-Soon Song: The Impact of Information and Communication Technologies (ICTs) on Rural Households. A Holistic Approach Applied to the Case of Lao People's Democratic Re- public. 2003.

Vol. 36 Daniela Lohlein: An Economic Analysis of Public Good Provision in Rural Russia. The Case of Education and Health Care. 2003.

Vol. 37 Johannes Woelcke. Bio-Economics of Sustainable Land Management in Uganda. 2003.

Vol. 38 Susanne M. Ziemek: The Economics of Volunteer Labor Supply. An Application to Countries of a Different Development Level. 2003.

Vol. 39 Doris Wiesmann: An International Nutrition Index. Concept and Analyses of Food Insecurity and Undernutrition at Country Levels. 2004.

Vol. 40 Isaac Osei-Akoto: The Economics of Rural Health Insurance. The Effects of Formal and Informal Risk-Sharing Schemes in Ghana. 2004.

Vol. 41 Yuansheng Jiang: Health Insurance Demand and Health Risk Management in Rural China. 2004.

Vol. 42 Roukayatou Zimmermann: Biotechnology and Value-added Traits in Food Crops: Relevance for Developing Countries and Economic Analyses. 2004.

Vol. 43 F. Markus Kaiser: Incentives in Community-based Health Insurance Schemes. 2004.

Vol. 44 Thomas Herzfeld: *Corruption begets Corruption.* Zur Dynamik und Persistenz der Korruption. 2004.

Vol. 45 Edilegnaw Wale Zegeye: The Economics of On-Farm Conservation of Crop Diversity in Ethiopia: Incentives, Attribute Preferences and Opportunity Costs of Maintaining Local Varieties of Crops. 2004.

Vol. 46 Adama Konseiga: Regional Integration Beyond the Traditional Trade Benefits: Labor Mobility contribution. The Case of Burkina Faso and Côte d'Ivoire. 2005.

Vol. 47 Beyene Tadesse Ferenji: The Impact of Policy Reform and Institutional Transformation on Agricultural Performance. An Economic Study of Ethiopian Agriculture. 2005.

Vol. 48 Sabine Daude: Agricultural Trade Liberalization in the WTO and Its Poverty Implications. A Study of Rural Households in Northern Vietnam. 2005.

Vol. 49 Kadir Osman Gyasi: Determinants of Success of Collective Action on Local Commons. An Empirical Analysis of Community-Based Irrigation Management in Northern Ghana. 2005.

Vol. 50 Borbala E. Balint: Determinants of Commercial Orientation and Sustainability of Agricultural Production of the Individual Farms in Romania. 2006.

Vol. 51 Pamela Marinda: Effects of Gender Inequality in Resource Ownership and Access on Household Welfare and Food Security in Kenya. A Case Study of West Pokot District. 2006.

Vol. 52 Charles Palmer: The Outcomes and their Determinants from Community-Company Contracting over Forest Use in Post-Decentralization Indonesia. 2006.

Vol. 53 Hardwick Tchale: Agricultural Policy and Soil Fertility Management in the Maize-based Smallholder Farming System in Malawi. 2006.

Vol. 54 John Kedi Mduma: Rural Off-Farm Employment and its Effects on Adoption of Labor Intensive Soil Conserving Measures in Tanzania. 2006.

Vol. 55 Mareike Meyn: The Impact of EU Free Trade Agreements on Economic Development and Regional Integration in Southern Africa. The Example of EU-SACU Trade Relations. 2006.

Vol. 56 Clemens Breisinger: Modelling Infrastructure Investments, Growth and Poverty Impact. A Two-Region Computable General Equilibrium Perspective on Vietnam. 2006.

Vol. 57 Meike Wollni: Coping with the Coffee Crisis. An Analysis of the Production and Marketing Performance of Coffee Farmers in Costa Rica. 2007.

Vol. 58 Franklin Simtowe: Performance and Impact of Microfinance. Evidence from Joint Liability Lending Programs in Malawi. 2008.

Vol. 59 Xiangping Jia: Credit Rationing and Institutional Constraint. Evidence from Rural China. 2008.

Vol. 60 Holger Seebens: The Economics of Gender and the Household in Developing Countries. 2009.

Vol. 61 Stephan Piotrowski: Land Property Rights and Natural Resource Use. An Analysis of Household Behavior in Rural China. 2009.

Vol. 62 Sebastian M. Scholz: Rural Development through Carbon Finance. Forestry Projects under the Clean Development Mechanism of the Kyoto Protocol. Assessing Smallholder Participation by Structural Equation Modeling. 2009.

Vol. 63 Jakob Rupert Friederichsen: Opening Up Knowledge Production through Participatory Research? Agricultural Research for Vietnam's Northern Uplands. 2009.

Vol. 64 Olivier Ecker: Economics of Micronutrient Malnutrition. The Demand for Nutrients in Sub-Saharan Africa. 2009.

Vol. 65 Julia Johannsen: Operational Assessment of Monetary Poverty by Proxy Means Tests. 2009

Vol. 66 Ephraim Nkonya / Nicolas Gerber / Philipp Baumgartner / Joachim von Braun / Alessandro De Pinto / Valerie Graw / Edward Kato / Julia Kloos / Teresa Walter: The Economics of Land Degradation. Toward an Integrated Global Assessment. 2011.

Vol. 67 S. Idriss Nazaire Houssou: Operational Poverty Targeting by Proxy Means Tests. Models and Policy Simulations for Malawi. 2013.

Vol. 68 Abdul Salam Lodhi: Education, Child Labor and Human Capital Formation in Selected Urban and Rural Settings of Pakistan. 2013.

Vol. 69 Evita Hanie Pangaribowo: Household Food Consumption, Women´s Asset and Food Policy in Indonesia. 2013.

Vol. 70 Dan Liu: China's New Rural Cooperative Medical Scheme. Evolution, Design and Impacts. 2013.

Vol. 71 Camille Saint-Macary: Microeconomic Impacts of Institutional Change in Vietnam's Northern Uplands. Empirical Studies on Social Capital, Land and Credit Institutions. 2014.

Vol. 72 Beatrice Wambui Muriithi: Commercialization of Smallholder Horticultural Farming in Kenya. Poverty, Gender, and Institutional Arrangements. 2014.

Vol. 73 Christian C. W. Grovermann: Assessment of Pesticide Use Reduction Strategies for Thai Highland Agriculture. Combining Econometrics and Agent-based Modelling. 2015.

Vol. 74 Dawit Diriba Guta: Bio-Based Energy, Rural Livelihoods and Energy Security in Ethiopia. 2015.

Vol. 75 Tigabu Degu Getahun: Industrial Clustering, Firm Performance and Employee Welfare. Evidence from the Shoe and Flower Cluster in Ethiopia. 2016.

Vol. 76 Abu Hayat Md. Saiful Islam: Impact of Technological Innovation on the Poor. Integrated Aquaculture-Agriculture in Bangladesh. 2016.

Vol. 77 Christine Husmann: The Private Sector and the Marginalized Poor. An Assessment of the Potential Role of Business in Reducing Poverty and Marginality in Rural Ethiopia. 2016.

Vol. 78 Qui Chen: Biomass Energy Economics and Rural Livelihood in Sichuan, China. 2018.

Vol. 79 Franziska Schünemann: Economy-Wide Policy Modeling of the Food-Energy-Water Nexus. Identifying Synergies and Tradeoffs on Food, Energy, and Water Security in Malawi. 2018.

www.peterlang.com

www.ingramcontent.com/pod-product-compliance
Ingram Content Group UK Ltd.
Pitfield, Milton Keynes, MK11 3LW, UK
UKHW021823290926
8980IPUK00004B/230

9 783631 756188